Remote Control

The Project Manager's Guide to Leading Teams,
Delivering Results, and Thriving from Anywhere

By Lynn R. Squire

Contents

Introduction 1

The Rise of Remote Work: How Project Management
Has Evolved 1

Why Remote Project Managers Must Develop New Skills
and Strategies... 2

What This Book Will Teach You: Practical Strategies,
Tools, and Stories 3

Who This Book Is For: Project Managers, Leaders, and
Anyone Managing Remote Teams 4

Part 1: The Fundamentals of Remote Project Management
.... 6

Chapter 1: The Remote Work Revolution............................ 7

The Shift To Remote And Hybrid Work Models.............. 8

Common Misconceptions About Remote Project
Management 10

Misconception #1: "Remote Workers Aren't as
Productive" 11

Misconception #2: "You Can Just Copy-Paste Office
Processes" 11

Misconception #3: "Too Many Tools Will Solve the
Problem" 12

Misconception #4: "Remote Work Kills Team Culture"
.... 13

Misconception #5: "You Need to Be Online All the
Time to Stay in Sync" 13

Misconception #6: "Remote Work Means Less Accountability".. 14

Busting Myths to Build Better Teams 14

Key Benefits and Challenges of Managing Remote Teams 15

The Benefits: Why Remote Teams Can Be a Game Changer 15

The Challenges: What Remote Teams Require You to Get Right 17

Balancing Both Worlds 19

Chapter 2: The Remote Project Manager's Mindset 21

The Importance of Adaptability, Communication, and Trust 22

Adaptability: The Superpower of Remote Leadership 22

Communication: Clear, Consistent, and Constructive 23

Trust: The Invisible Glue 24

Managing Outcomes Over Micromanaging Tasks 25

Focus on Results, Not Clock-Watching 26

Create a Culture of Ownership 26

Build Accountability Without Hovering 27

Let Go to Lead Better 28

Staying Productive and Avoiding Burnout as a Remote Leader.. 28

Productivity Isn't About Being Busy.......................... 29

Build a Sustainable Routine (That Actually Works) .. 30

Say No More Often 30

Watch for the Red Flags.. 31

You Can't Pour From an Empty Coffee Cup.............. 31

Chapter 3: Setting Up Your Virtual Team for Success 33

How to Build a High-Performing Remote Project Team 34

Start With the Right Mix of Skills and Roles 34

Define the "Why" From Day One............................. 35

Establish Norms and Rituals Early............................ 35

Use Roles, Not Job Titles, to Clarify Ownership........ 36

Hire for Initiative, Not Just Execution 36

Hiring and Onboarding Remote Team Members
Effectively 37

What Makes Remote Hiring Different? 37

Onboarding: The Make-or-Break Phase 38

Establishing Clear Roles, Responsibilities, and
Expectations 41

Why Clarity Matters More Remotely....................... 42

Define Roles Beyond Job Titles 42

Set Expectations Around Communication and
Availability.. 43

Co-Create the Rules (Don't Just Dictate Them) 44

Document It All (Then Actually Use It)..................... 45

Part 2: Managing Projects in a Remote World 46

Chapter 4: Remote Project Planning and Execution......... 47

Adjusting Traditional Project Methodologies for Remote Teams 48

Waterfall: The Classic Blueprint (and How to Modernize It Remotely) 48

Agile: Flexibility at Its Finest (With Remote-First Tweaks) 49

Scrum: Rituals With Remote Rigor 50

Choosing the Right Method (or Mix) 51

Defining Clear Goals, Deliverables, and Accountability Structures 52

Define Goals That Actually Mean Something 52

Clarify Deliverables (Don't Assume Anything) 53

Build Accountability Without Hovering 54

Best Practices for Virtual Project Kickoff Meetings 55

1. Know the Purpose of Your Kickoff 56

2. Send a Pre-Kickoff Brief 56

3. Curate Your Attendees Thoughtfully 57

4. Structure the Agenda Around Clarity and Connection 58

5. Set the Tone – Literally 59

6. Use Visuals to Anchor Attention 60

7. Capture and Share the Kickoff Summary 60

8. Watch for Early Warning Signs 61

Virtual Kickoff = Culture in Action 61

Chapter 5: Communication Strategies for Remote Teams 63

Choosing the Right Tools: Slack, Microsoft Teams, Zoom, Email, and More 64

Slack: The Digital Office Hallway 64

Microsoft Teams: The Enterprise Communication Hub 65

Zoom: The Face-to-Face Stand-In 66

Email: The Digital Paper Trail 68

Project Management Tools: The Source of Truth 69

Other Tools Worth Mentioning 70

Build a Communication Stack, Not a Frankenstein .. 70

Asynchronous vs. Synchronous Communication: When to Use Each 71

The Case for Asynchronous Communication 72

The Case for Synchronous Communication 74

Choosing the Right Mode 75

Building a Culture That Supports Both 76

Examples in Practice 76

Common Pitfalls to Avoid 77

Avoiding Miscommunication and Fostering Transparency 78

Prioritize Clarity in Every Message 79

Use Shared Docs and Templates to Eliminate Guesswork 80

Make Statuses and Progress Visible 81

Establish a "Default to Documentation" Culture 82

Create Communication Norms (and Write Them Down).. 82

Foster a Feedback Culture 83

Lean on Visual and Async Tools for Clarity 84

Build Transparency Into Your Rituals........................ 85

Chapter 6: Building Trust and Team Culture from a Distance... 87

Encouraging Engagement and Motivation Remotely ... 88

Understand What Drives Each Person...................... 88

Use Recognition and Celebration 89

Make Work Visible 89

Provide Growth and Learning Opportunities 89

Use Pulse Surveys to Listen Actively 90

Align Work to Purpose 90

Encourage Social Connection................................... 91

Set Clear Expectations 91

Provide Flexibility with Boundaries 92

Lead by Example 92

Virtual Team-Building Activities That Actually Work 92

Virtual Coffee Chats or "Donut" Pairings................. 93

Virtual Game Sessions (That Aren't Cheesy) 93

Shared Learning Experiences 94

Recognition Rituals 94

Virtual Wellness Challenges..................................... 94

Storytelling Sessions 95

Cross-Team Projects 95

Handling Conflicts and Difficult Conversations Remotely
.... 96

Recognizing Early Warning Signs 96

Preparing for Difficult Conversations 97

Creating a Safe and Focused Space 97

Listening Actively and Navigating Emotions 98

Following Up and Documenting Agreements........... 98

Building a Long-Term Conflict-Resilient Culture....... 99

Part 3: Overcoming Remote Work Challenges 101

Chapter 7: Managing Productivity and Performance 102

Setting SMART Goals for Remote Teams 103

Why SMART Goals Work So Well Remotely 103

Breaking Down SMART Goals 104

How to Set SMART Goals Remotely........................ 106

SMART Goals in Action: Three Remote Role Examples
.... 107

Common Mistakes to Avoid................................... 108

Measuring Performance Without Micromanaging..... 109

Why Micromanagement Doesn't Work (Especially
Remotely)... 109

Shift from Time-Based to Outcome-Based
Management... 110

Set Clear Performance Metrics............................. 111

Tools for Transparent Performance Tracking 112

Create a Regular Feedback Rhythm 113

Recognize and Reward Results 114

Support Underperformance with Coaching, Not
Control 114

Encourage Self-Assessment and Ownership 115

Overcoming Distractions and Time Zone Differences 116

Tackling Distractions: The Home Office Productivity
Killers 116

Managing Time Zone Differences Like a Pro 118

Chapter 8: Handling Remote Project Risks and Issues ... 121

Identifying Risks Unique to Remote Projects 122

1. Communication Gaps and Message Misfires 122

2. Tech Dependencies and Platform Failures 123

3. Team Disengagement or Isolation 124

4. Time Zone Silos and Collaboration Delays 124

5. Lack of Visibility into Progress 125

6. Decision-Making Bottlenecks 126

Contingency Planning for Communication Breakdowns,
Tech Failures, and Disengagement 127

1. Communication Breakdowns 127

2. Tech Failures 129

3. Disengagement and Team Silence 130

How to Maintain Resilience When Things Go Wrong. 132

1. Normalize the Bumps in the Road 132

2. Use Post-Mortems as Learning Tools 133

3. Communicate with Transparency and Empathy. 133

4. Protect Energy and Morale 134

Chapter 9: Running Effective Remote Meetings 136

How to Eliminate Zoom Fatigue and Wasted Meeting
Time.. 137

Meeting Best Practices: When, How, and Why to Meet
.... 138

1. Always Know the "Why" 138

2. Use an Agenda – Always 139

3. Invite Only the Necessary People 139

4. Establish Ground Rules 139

5. Use the Right Meeting Type 140

6. Timebox Everything 140

7. Start and End on Time... 140

8. End with Clear Outcomes 140

9. Consider Async Alternatives 141

10. Review and Improve Regularly 141

Using AI, Recordings, and Summaries to Improve
Meeting Efficiency... 142

1. Recordings Are Your Asynchronous Superpower 142

2. Let AI Handle the Note-Taking............................ 143

3. Use Summaries to Replace Recaps and Repeats 144

4. Keep a Knowledge Trail...................................... 144

5. Balance Efficiency With Human Connection 145

Part 4: Advanced Strategies for Remote Project Success146

Chapter 10: Leveraging Technology to Stay Ahead 147

AI and Automation Tools That Make Remote Project
Management Easier 148

1. ClickUp AI 148

2. Fireflies.ai 149

3. Motion............................. 150

4. Otter.ai 150

5. Loom AI (Async Video Messaging) 151

6. Trello Automation (Butler)............................... 151

7. Notion AI 152

8. Zapier (Automation Integration Platform).......... 153

9. Reclaim.ai 154

10. Hypercontext 154

11. Jira Smart Automation & AI Features 155

12. Asana AI & Automation.................................. 156

Integrating Project Management Platforms (Jira, Trello,
Asana, ClickUp)... 157

Why Integration Matters for Remote Teams 157

Jira Integration Highlights 157

Trello Integration Highlights 158

Asana Integration Highlights.................... 159

ClickUp Integration Highlights 159

Choosing and Connecting the Right Platform......... 160

Cybersecurity Best Practices for Remote Teams 161

    1. Start with a Cybersecurity Culture...................... 161

    2. Use Strong Passwords and Two-Factor
    Authentication (2FA)... 162

    3. Secure Devices and Connections 162

    4. Control Access to Tools and Data 163

    5. Use Secure File Sharing and Storage.................. 163

    6. Prevent Phishing and Social Engineering............ 164

    7. Create a Security Playbook for Remote Teams .. 165

    8. Audit Regularly and Update Policies................... 165

    9. Balance Security with Usability........................... 166

Chapter 11: Leading Remote Agile & Hybrid Teams....... 167

Adapting Agile Principles to Remote Work................. 168

    1. Customer satisfaction through early and
    continuous delivery of valuable software 168

    2. Welcome changing requirements, even late in
    development... 169

    3. Deliver working software frequently, with a
    preference for shorter timescales 169

    4. Business people and developers must work
    together daily throughout the project 169

    5. Build projects around motivated individuals. Give
    them the environment and support they need and
    trust them to get the job done 170

6. The most efficient and effective method of conveying information to and within a development team is face-to-face conversation 170

7. Working software is the primary measure of progress.. 171

8. Agile processes promote sustainable development. The sponsors, developers, and users should be able to maintain a constant pace indefinitely 171

9. Continuous attention to technical excellence and good design enhances agility 171

10. Simplicity – the art of maximizing the amount of work not done – is essential 172

11. The best architectures, requirements, and designs emerge from self-organizing teams........................ 172

12. At regular intervals, the team reflects on how to become more effective, then tunes and adjusts its behavior accordingly... 173

Running Remote Sprints, Stand-Ups, and Retrospectives 173

Remote Sprint Planning 174

Remote Daily Stand-Ups 175

Remote Sprint Reviews / Demos 176

Remote Sprint Retrospectives 178

Additional Tips for Remote Agile Rituals 179

Case Studies of Successful Remote Agile Teams......... 179

GitLab: Pioneering a Fully Remote Agile Model 180

Synchrony Financial: Rapid Agile Response to Employee Needs 181

Zapier: Scaling Agile in a Remote-First Environment 182

Agile Anywhere – Leading with Flexibility and Focus . 183

Chapter 12: The Future of Remote Project Management 185

Trends Shaping the Future of Remote Work 186

1. AI and Automation Everywhere........................... 186

2. The Rise of Async-First Cultures........................... 187

3. Work-Life Integration and Mental Health Prioritization 187

4. Global Talent, Local Challenges 188

5. Outcome-Based Performance Management...... 188

6. The Maturation of Hybrid Models 189

7. Increased Focus on Security and Digital Literacy 189

8. Project Managers as Coaches, Not Controllers .. 190

9. The Virtual Office is Getting Smarter.................. 190

Hybrid Work Models: Finding the Right Balance 191

Define the Why of In-Person Time 191

Normalize Remote Participation as the Default 192

Invest in Asynchronous Collaboration 192

Design for Equity, Not Uniformity 193

Create Clear, Flexible Policies 193

Rethink Office Design and Purpose 193

Measure What Matters 194

Continuous Improvement as a Remote Project Manager
.... 194

   1. Retrospectives Aren't Just for Teams 195

   2. Develop Your Remote-Specific Leadership Skills 195

   3. Track What You Want to Improve 196

   4. Build a Feedback-Rich Culture 197

   5. Share What You're Learning 197

   6. Protect Time for Strategic Thinking 198

Conclusion & Final Takeaways 199

Recap of Key Lessons 199

   1. The Remote Landscape Has Changed, And So Must
You 199

   2. Mindset Is Your Foundation 199

   3. Team Setup Is Everything.................................. 200

   4. Planning Still Matters, It Just Looks Different..... 200

   5. Communication Is Your Superpower 200

   6. Culture Doesn't Happen by Accident................. 201

   7. Productivity Is About Outcomes, Not Hours....... 201

   8. Risk and Resilience Go Hand in Hand 201

   9. Meetings Must Be Intentional 201

   10. Technology Is Your Teammate......................... 202

   11. Agile Can Work Anywhere 202

   12. The Future Is Hybrid, Flexible, and Human-
Centered 202

13. Your Growth Is the Team's Growth 202

A Personal Challenge: Taking Action Today 203

Resources for Continued Learning 205

Books 205

Podcasts 206

Online Courses & Certifications 206

Additional Guides 207

Introduction

Welcome to the wild world of remote project management. If you've ever stared at a blinking Zoom cursor waiting for a response, sent a message into the void of Slack never to be seen again, or wondered whether your team members are hard at work or just perfecting their sourdough recipes, congratulations – you're in the right place.

Managing projects remotely isn't just about working from the comfort of your home in pajama pants (though that's a nice perk). It's about mastering the art of getting things done when your team is scattered across time zones, juggling multiple responsibilities, and possibly using the world's slowest Wi-Fi connection. Traditional project management was already tough, but remote work adds an extra layer of complexity – one that requires new strategies, tools, and ways of thinking.

The Rise of Remote Work: How Project Management Has Evolved

It wasn't that long ago that most project managers could walk over to someone's desk, tap them on the shoulder, and ask, "Hey, where's that report?" Today, you're lucky if your message doesn't get buried under 27 unread Slack notifications.

Remote work has exploded in recent years, fueled by advancements in technology and accelerated by global events that forced companies to rethink the way work gets done. What started as a temporary shift has now become the norm for many businesses. Companies have realized

1

that remote teams can be just as productive – if not more so – than their in-office counterparts. The challenge is that traditional project management approaches weren't designed with remote teams in mind.

That's where this book comes in. We're not here to dwell on the difficulties of remote work (we all know them too well). Instead, we'll focus on solutions – practical, battle-tested strategies to help you lead with confidence, keep your projects on track, and avoid the chaos that comes from miscommunication, lack of visibility, and the dreaded "Can everyone see my screen?" moments.

Why Remote Project Managers Must Develop New Skills and Strategies

Managing a project in a remote environment isn't as simple as copying and pasting office-based strategies into a virtual setting. It requires a shift in mindset and the development of new skills.

For starters, communication is no longer as simple as walking into a conference room. Without the luxury of in-person interactions, remote project managers need to be crystal clear in their messaging, setting expectations, and ensuring nothing gets lost in translation.

Trust is another crucial factor. In an office setting, it's easy to build trust through casual interactions – grabbing coffee, small talk, or spontaneous brainstorming sessions. In a remote environment, project managers must be intentional about fostering trust, ensuring accountability without resorting to micromanagement, and creating a team culture where people feel connected despite the distance.

And let's not forget about technology. The right tools can make or break a remote team's efficiency. But with so many options available – Slack, Microsoft Teams, Asana, Trello, Jira, Zoom, Notion, and the list goes on – it can be overwhelming to figure out what works best. This book will help you navigate the ever-growing sea of tech options and find the right mix for your team.

What This Book Will Teach You: Practical Strategies, Tools, and Stories

This isn't a book filled with abstract theories that sound great on paper but don't work in real life. Everything here is designed to be immediately useful – something you can apply to your projects today.

Throughout these pages, you'll learn how to:

- Set up your remote team for success from day one

- Adapt project methodologies (Agile, Scrum, Waterfall) for a remote setting

- Use the right communication tools effectively (and avoid tool overload)

- Build a strong remote team culture and maintain high engagement

- Overcome common remote work challenges, from time zone issues to miscommunication

- Run meetings that don't make everyone wish they had scheduled a dentist appointment instead

3

- Leverage technology and automation to improve project efficiency

You don't need a formal project management title to benefit from this book. If you're responsible for leading a team – whether you're a project manager, program manager, entrepreneur, or just the "go-to" person who keeps things running – you'll find valuable insights here.

This book is for:

- Experienced project managers looking to refine their remote leadership skills

- Business leaders and entrepreneurs managing distributed teams

- New managers navigating the world of remote work for the first time

- Anyone frustrated with ineffective remote collaboration and seeking practical solutions

If you've ever struggled with unresponsive team members, meetings that go nowhere, or unclear project expectations, you're not alone. The good news? There are proven strategies to make remote project management smoother, more efficient, and – dare we say – more enjoyable.

So, are you ready to master remote project management?

By the time you finish this book, you'll have the knowledge, tools, and confidence to lead remote projects successfully. You'll learn how to communicate effectively, manage performance without micromanaging, keep your team engaged, and overcome the challenges that come with working from different locations.

So, grab your favorite coffee, get comfortable, and let's dive in. Remote project management doesn't have to be a struggle – with the right approach, it can be one of the most rewarding ways to lead a team. Let's get started.

Part 1: The Fundamentals of Remote Project Management

Chapter 1: The Remote Work Revolution

There was a time when "working from home" meant someone was waiting for a furniture delivery or quietly binge-watching daytime TV while answering the occasional email. Fast forward a few years, and now entire industries are built on the idea that you can lead a global team from your living room (and yes, possibly with a laundry basket just out of frame).

Remote work isn't a trend anymore – it's a fundamental shift in how we define the workplace. The cubicle has been replaced by the kitchen table, the water-cooler chat has gone digital, and "commute time" is now how long it takes your laptop to boot up. This change didn't come with a manual, and project managers everywhere have been figuring it out in real time – often while muting a barking dog, dodging background chaos, or explaining once again that yes, this *is* their office.

The remote work revolution has changed more than just our settings. It's rewritten the rules of collaboration, redefined productivity, and forced us to reconsider what leadership looks like when no one's sitting across the conference table. For project managers, it's been both an exciting opportunity and a bit like being handed a puzzle with no picture on the box.

But here's the thing: we're not just surviving this shift – we're shaping it. This chapter is your starting point for understanding how remote work got here, why it's sticking around, and what it means for the way you manage, lead, and get results in a world that no longer fits neatly into four walls.

The Shift To Remote And Hybrid Work Models

It started as a workaround. A temporary fix. A *"we'll figure it out for now"* kind of solution.

Before 2020, working remotely was often seen as a special perk – something reserved for freelancers, digital nomads, or that one person in the office who "only came in on Thursdays." Most companies believed that real work happened in real offices with real desks and real fluorescent lighting. Then came a global shake-up that flipped that belief on its head, pushed us all out of the office, and said, "Let's see how you manage *now*."

And manage we did – sometimes awkwardly, sometimes brilliantly, but always with a fast-forward button on innovation. As offices shut their doors and employees logged in from kitchen tables, home offices, and sometimes creatively propped-up ironing boards, something unexpected happened: work kept happening.

What began as a global emergency response slowly revealed something much more transformative. Productivity didn't crash. Deadlines didn't vanish into thin air. In fact, many teams found themselves thriving without the usual distractions of office life. No more hour-long commutes, impromptu meetings that could have been emails, or struggling to find a quiet corner to concentrate. Teams found out they could collaborate from different cities, states, and even continents. The workplace, as we knew it, had been permanently altered.

Remote work quickly evolved from an emergency plan to a viable – sometimes preferable – model. And as organizations watched employees get the job done from home, it became clear that this wasn't just a temporary blip. It was a shift. A major one.

Of course, not every job could go remote. And not every employee wanted to work from their living room forever. That's where hybrid models entered the scene, offering a compromise between the structure of the office and the flexibility of remote life. These models range from highly structured setups – like rotating in-office schedules – to looser approaches where employees decide when and how often to come in. The idea is to give people the best of both worlds: the ability to collaborate in person when it matters, and the freedom to focus independently when it doesn't.

But here's where things get interesting – and a bit complicated for project managers.

No two remote or hybrid models are exactly the same. Some teams are spread across multiple time zones. Others might have half the team in the office and the other half tuning in virtually. Some organizations embrace full flexibility, while others expect employees to follow a set rhythm. It's a spectrum, and navigating it requires a thoughtful approach.

This new world of work impacts everything: how teams communicate, how progress is tracked, how trust is built, and how culture is sustained. As a project manager, your job now includes understanding the rhythm of your specific work model and leading in a way that supports both collaboration and autonomy – no matter where your team is logging in from.

The shift to remote and hybrid work isn't just a logistical change. It's a cultural one. It challenges the assumption that productivity is linked to proximity and invites us to rethink what effective teamwork looks like. Instead of managing by presence ("They're at their desk, so they must be working"), you now lead by outcomes. Instead of daily check-ins in person, you build systems that keep everyone aligned, engaged, and moving forward – even when they're moving at slightly different paces from different places.

This chapter, and this book as a whole, will help you understand how to make that shift – not just in tools or workflows, but in mindset and strategy. Because the future of work isn't coming. It's already here. And it's remote, hybrid, and full of opportunity for those willing to adapt.

Common Misconceptions About Remote Project Management

Remote project management has come a long way – but let's be honest, it's still battling a few outdated myths. These misconceptions can hold teams back, frustrate leaders, and lead to inefficient practices that do more harm than good. If you've ever heard someone say, "You can't *really* manage a project without being in the same room," then you know exactly what we're talking about.

Let's set the record straight.

This one's been floating around for years – usually whispered in offices with a suspicious side-eye toward anyone who dared work from home. The idea that remote employees are secretly slacking off, binge-watching TV, or doing laundry instead of working has been surprisingly persistent.

But research tells a different story. Numerous studies – including long-term ones from Stanford, Harvard Business Review, and Gallup – have shown that remote workers can be *more* productive than their in-office counterparts, especially when given clear goals and the autonomy to manage their time.

The key is trust and structure. When team members understand what's expected of them and are given the tools to succeed, location becomes almost irrelevant. Remote project management isn't about hovering – it's about setting the stage for success and then stepping back to let your team perform.

Misconception #2: "You Can Just Copy-Paste Office Processes"

Spoiler alert: you can't. What works in a co-located office doesn't always translate well to a distributed environment. From communication norms to task delegation and performance check-ins, many office-based habits need a full makeover in a remote context.

For instance, informal hallway chats that once served as impromptu status updates now need to be replaced with structured touchpoints. Team visibility isn't about who's sitting at their desk — it's about how clearly tasks and responsibilities are tracked, shared, and communicated.

Trying to replicate office dynamics online without adjustment is a fast track to miscommunication, confusion, and frustration. Remote project management calls for rethinking — not replicating — how your team works together.

Misconception #3: "Too Many Tools Will Solve the Problem"

When teams go remote, there's a natural instinct to compensate with tools. "Just add another app," they say. And before long, your team is toggling between six platforms just to post a status update.

While technology is critical for remote collaboration, more tools don't automatically mean better communication. In fact, tool overload can create more problems than it solves — scattered conversations, missed updates, and constant context-switching that drains focus.

Remote project managers must be intentional about choosing the right tools, establishing norms for how (and when) to use them, and avoiding the all-too-common trap of digital clutter. The focus should always be on clarity and simplicity — not chasing the next shiny platform.

Here's the truth: remote work doesn't kill culture. Neglect does.

It's entirely possible to have a vibrant, engaged team culture in a remote environment. It just requires more intentionality. Culture isn't built by proximity – it's built by shared values, clear communication, mutual trust, and moments of connection, even if they happen over a screen.

Remote project managers can nurture culture by creating space for informal check-ins, celebrating wins, encouraging collaboration, and promoting psychological safety. The best remote cultures aren't accidental – they're engineered, just like a strong project plan.

One of the biggest traps remote teams fall into is the belief that everyone has to be available at all hours to stay aligned. This "always-on" mindset leads to burnout, blurred boundaries, and constant interruptions.

In reality, asynchronous communication – where people contribute on their own schedule – is one of the most powerful advantages of remote work. It allows for flexibility, deeper focus time, and respect for time zones and work styles.

Great remote project management leans into async tools and strategies, building workflows that don't rely on everyone being online at the same moment. Being "in sync" is less about simultaneous presence and more about structured communication, clear documentation, and trust in the process.

Misconception #6: "Remote Work Means Less Accountability"

Let's flip that: remote work requires *more* accountability, not less. Without physical proximity, project managers can no longer rely on visibility as a measure of engagement. Instead, success depends on clarity: clear expectations, clear goals, and clear check-ins.

Accountability in a remote setting thrives when people understand their roles, know how their work connects to the bigger picture, and are empowered to own their outcomes. Micromanagement becomes not only unsustainable, it becomes counterproductive. The best remote teams operate on alignment and autonomy, not surveillance.

Busting Myths to Build Better Teams

These misconceptions aren't just misunderstandings — they're obstacles. They affect how teams are structured, how success is measured, and how people feel about their work. As a remote project manager, part of your job is to challenge these outdated ideas and lead with clarity, creativity, and confidence.

Because here's the truth: remote project management isn't second-tier. It's not "less than." It's just different. And when done right, it's every bit as powerful – if not more so – than traditional models.

So go ahead: ditch the myths, rethink the playbook, and lead your team with strategies that actually match how people work today.

Key Benefits and Challenges of Managing Remote Teams

Managing a remote team can feel like unlocking a cheat code for productivity and flexibility – or, depending on the day, like juggling flaming swords over a Wi-Fi connection that just decided to take the afternoon off. It's a dynamic experience, filled with both incredible advantages and unique obstacles.

Understanding both sides of the coin will help you prepare, adapt, and lead more effectively, turning potential headaches into strategic wins and setting your team up for long-term success.

The Benefits: Why Remote Teams Can Be a Game Changer

1. *Access to a Broader Talent Pool*

When geography is no longer a barrier, your hiring options open up dramatically. Need a brilliant developer, designer, or data analyst? You're no longer limited to who's within commuting distance. You can hire the best person for the role, whether they're across town or across the globe.

This global talent pool increases your ability to build diverse, innovative teams with a wide range of perspectives. It also helps fill niche or highly specialized roles that might be harder to recruit for locally.

2. *Flexibility Increases Satisfaction and Retention*

Numerous studies — from companies like Buffer, Gallup, and Owl Labs — show that employees value flexibility as much as (and sometimes more than) salary or title. Remote work often allows team members to design their workdays around their most productive hours, personal obligations, and natural rhythms.

This autonomy leads to higher job satisfaction, which in turn boosts morale and retention. Happy team members tend to stick around longer, work more creatively, and bring a better attitude to their daily responsibilities.

3. *Fewer Distractions, More Deep Work*

Open-plan offices were supposed to foster collaboration. In reality, they often led to noise, interruptions, and the dreaded "drive-by" requests. Remote work allows individuals to create their own focused environments, reducing distractions and increasing opportunities for deep, uninterrupted work.

That said, it's not automatic; leaders must still help team members establish boundaries and healthy habits to maintain this focus (more on that in later chapters).

4. Cost Savings for Everyone

Remote work can lead to significant savings, not just for the company, but also for employees. Companies can reduce overhead costs tied to office space, utilities, equipment, and supplies. Meanwhile, team members save on commuting, lunches out, wardrobe costs, and sometimes even childcare.

Some businesses choose to reinvest these savings into better tools, professional development, or team perks, all of which can strengthen team performance and engagement.

The Challenges: What Remote Teams Require You to Get Right

Remote work isn't without its friction points. Some are logistical, others cultural, and a few are purely human. The key is recognizing them early and addressing them with clear, intentional strategies.

1. Communication Gaps

Without in-person cues – body language, tone, spontaneous side conversations – misunderstandings can multiply quickly. A short message that seemed efficient to one person might come off as curt or unclear to another.

Remote project managers must be proactive about communication – choosing the right tools, setting expectations around tone and clarity, and encouraging frequent check-ins without overwhelming the team with nonstop messages.

2. Time Zone Coordination

Working across multiple time zones is one of the trickiest remote challenges. Scheduling meetings becomes a game of calendar Tetris, and delays can stack up when people work on different clocks.

While this isn't insurmountable, it does require a balance of asynchronous processes and respectful scheduling. Leaders who thrive in remote environments understand how to plan around these time differences and ensure everyone still feels part of the same team.

3. Building and Sustaining Trust

In a remote world, you can't rely on casual run-ins or after-meeting chit-chat to build rapport. Trust must be cultivated intentionally through reliability, empathy, and transparency.

When trust is missing, people start second-guessing each other. Deadlines get missed, responsibilities fall through the cracks, and collaboration suffers. A strong remote leader creates an environment where people feel seen, supported, and accountable.

4. Visibility and Performance Tracking

In an office, you can often "see" who's working, though let's be honest, desk presence doesn't always equal productivity. In a remote setting, visibility shifts from physical presence to performance outcomes.

This requires better systems for tracking progress, providing feedback, and celebrating wins. Remote project managers must focus on what's being accomplished, not how many hours someone is logged into Slack.

5. Isolation and Burnout

Ironically, working from home can make people feel *too* alone. Without natural social interaction or a commute to signal the start and end of the day, some remote workers struggle with overworking, loneliness, or feeling disconnected from the team.

Recognizing the signs of burnout, creating social touchpoints, and promoting a healthy work-life balance are all essential parts of remote leadership. Your people are still your people... even if they're working in bunny slippers.

Balancing Both Worlds

Every benefit of remote work comes with a corresponding challenge and vice versa. The key to success lies in intentional leadership: creating systems that support productivity, encouraging open communication, and designing a work culture that values results over appearances.

Remote project management isn't about perfection, it's about constant improvement. With the right mindset and a clear strategy, the advantages of managing a remote team can far outweigh the obstacles.

Chapter 2: The Remote Project Manager's Mindset

You've got the tools. You've got the team. You've even managed to get everyone into the same virtual meeting at the same time – a rare planetary alignment. But here's the thing: none of that matters if your mindset isn't right.

Remote project management isn't just a shift in logistics; it's a shift in leadership philosophy. You're not just moving your processes online – you're reimagining what it means to lead when there's no office, no hallway conversations, and no visual cues to fall back on. The rules have changed, and so have the expectations.

Being a great remote project manager means letting go of some deeply ingrained habits – like measuring productivity by desk time or equating busyness with effectiveness. It means embracing a leadership style that prioritizes outcomes over optics, trust over control, and clarity over chaos. That can be a challenge, especially if you've spent most of your career managing in person.

This chapter is about retooling the way you think so you can lead your team with confidence, even when you can't see what they're doing every minute of the day. Because successful remote leadership doesn't come from having all the answers – it comes from having the right mindset to ask the right questions, set the right tone, and build the kind of team that doesn't just survive remotely but actually thrives.

The Importance of Adaptability, Communication, and Trust

Remote project management is less about having a perfect plan and more about knowing how to adjust when that plan inevitably changes – probably mid-Zoom call while someone's dog is barking in the background. In this environment, three qualities rise above the rest: adaptability, communication, and trust. Together, they form the backbone of effective remote leadership. Without them, even the best strategies can fall flat.

Adaptability: The Superpower of Remote Leadership

In a remote setting, rigid leadership doesn't just slow things down – it breaks things. Project scopes shift, priorities change, and team members may face unpredictable challenges (technical issues, personal distractions, or even just that one app that refuses to sync properly).

Being adaptable means responding to these changes with flexibility instead of frustration. It's the ability to pivot when timelines shift, rethink workflows when tools aren't working, and adjust expectations without sacrificing quality. It's about staying steady when the plan doesn't go as planned, which, let's be honest, is often.

Research from the Project Management Institute (PMI) highlights that the most effective project managers are those who demonstrate high levels of agility – able to move between methodologies, shift strategies, and course-correct without losing momentum. In a remote world, this skill is essential.

Adaptability in Action Might Look Like:

- Reassigning tasks quickly when someone is unexpectedly offline.
- Switching to an asynchronous process to accommodate different time zones.
- Reprioritizing deliverables when a client's needs suddenly change.

It's not about being reactive. It's about being responsive – with purpose and calm.

Communication: Clear, Consistent, and Constructive

If adaptability keeps your team moving, communication keeps everyone from running in different directions.

When you're managing a team you can't physically see, every message counts. Vague instructions become misunderstandings. Missed updates become missed deadlines. And silence? That becomes a vacuum where assumptions and confusion take over.

Good remote communication is:

- **Clear**: Say what you mean, and avoid unnecessary jargon or assumptions.
- **Consistent**: Keep updates flowing on a regular schedule. Don't leave your team guessing.
- **Constructive**: Feedback should be specific and helpful, not vague or critical.

It also means choosing the right channel. Not every conversation needs a meeting, just like not every question belongs in a group thread. Knowing when to use Slack, when to send an email, and when to jump on a quick call is a critical skill; your team will thank you for not turning every task update into a 30-minute video chat.

Studies show that remote teams with strong communication practices are more productive, more engaged, and less likely to suffer from confusion or burnout. The goal isn't just to communicate more, it's to communicate better.

Trust: The Invisible Glue

You can't lead a remote team effectively if you don't trust your people and if they don't trust you back. In the absence of face-to-face interaction, trust becomes your most valuable currency.

But here's the twist: trust looks a little different in remote environments. It's not built through casual office banter or Friday pizza parties. It's built through reliability, transparency, and mutual respect.

For managers, that means:

- Giving people autonomy instead of micromanaging.
- Being upfront about changes, challenges, and expectations.
- Following through on promises and being available when your team needs you.

For team members, it means:

- Owning their work and delivering on commitments.
- Communicating proactively when something's unclear or off-track.
- Showing integrity, even when no one's watching.

The Harvard Business Review reports that high-trust organizations outperform low-trust ones in everything from engagement to innovation. In remote teams, where psychological safety and independence are even more important, trust isn't a soft skill, it's a strategic advantage.

Managing Outcomes Over Micromanaging Tasks

Let's address the elephant in the virtual room: remote project managers who try to control every detail of their team's work are not managing – they're just playing an elaborate (and exhausting) game of digital Whac-A-Mole.

In an office setting, it's tempting to equate visibility with control. You could glance across the room, see who was typing, who looked busy, who always had their headphones on – and assume productivity from there. But remote work strips away those visual cues, and for some managers, that's terrifying. Suddenly, they can't "see" the work happening. So they overcompensate with constant pings, endless check-ins, and detailed task lists that read like IKEA manuals.

Here's the reality: micromanagement doesn't scale – especially not remotely. It breeds resentment, stifles creativity, and sends the message that you don't trust your team. And if your team senses that you're hovering (even digitally), they'll either disengage or burn out trying to prove themselves.

That's why great remote leadership is all about managing outcomes, not minute-to-minute activity.

Focus on Results, Not Clock-Watching

In a distributed team, what matters isn't how long someone was "green" on Slack – it's what they delivered. Remote project managers must shift their focus from time spent working to value created. What did your team accomplish today? What milestones moved forward? What problems were solved?

Define clear outcomes and let your team determine the best path to get there. You might be surprised: when people are empowered to own their results, they often exceed expectations – and they do it in ways you wouldn't have thought of.

This shift requires a few core practices:

- **Set crystal-clear expectations.** What does "done" look like? When is it due? Who owns it?

- **Agree on how progress is reported.** Whether it's a weekly update, a project board, or a shared dashboard – make visibility automatic.

- **Prioritize outputs over optics.** Don't reward "looking busy" – reward moving the needle.

Create a Culture of Ownership

Micromanagement is often a symptom of unclear roles or shaky trust. If you want your team to take ownership, they need to know what's theirs to own – and they need the space to do it.

That means:

- Delegating with intent: assign real responsibility, not just tasks.

- Encouraging autonomy: let team members solve problems in their own way.

- Supporting decision-making: don't be the bottleneck – be the coach.

Ownership doesn't mean letting go entirely. It means shifting from "Do this exactly my way" to "Here's the outcome we want – how do you want to approach it?"

Pro tip: you'll still need to check in – but check in on progress and blockers, not whether they've formatted the spreadsheet the way you like it. (Unless you're writing a book on spreadsheets. In which case, carry on.)

Build Accountability Without Hovering

One of the trickiest parts of remote work is creating a system of accountability that doesn't feel like surveillance. Nobody wants to be tracked like a pizza delivery. And honestly, if your accountability system depends on constant observation, it's already broken.

Here's what works better:

- **Mutual agreements:** Set expectations together and agree on how success will be measured.

- **Transparent workflows:** Use project management tools to make progress visible without needing to ask for updates constantly.

- **Regular reviews:** Weekly meetings, one-on-one check-ins, and team demos help everyone reflect, realign, and celebrate progress.

Accountability should feel like momentum, not micromanagement. The goal is to make it easier to stay on track — not harder to breathe.

Let Go to Lead Better

In remote leadership, letting go isn't a weakness — it's a strength. Your job isn't to manage every task. It's to create an environment where great work can happen without you needing to oversee every keystroke.

Trust your team. Empower them. Give them the tools, clarity, and autonomy they need — and then step back. You'll be amazed at what they can achieve when you stop managing like you're standing over their shoulder and start leading like you're standing beside them.

Staying Productive and Avoiding Burnout as a Remote Leader

Let's be real: remote leadership can feel like running a marathon while juggling flaming torches... on a unicycle... during a Zoom call. You're managing projects, supporting your team, answering messages at all hours, and somehow still trying to eat a warm lunch. If you've ever felt like the lines between "work" and "life" have completely vanished — you're not alone.

While remote work offers freedom and flexibility, it also comes with an undercurrent of danger: burnout. And here's the kicker — remote leaders are often the first to hit the wall. Why? Because they're constantly "on," trying to hold everything together, and often feel guilty stepping away when their team is depending on them.

The good news? It doesn't have to be this way. Staying productive *and* sane as a remote leader isn't a mythical dream – it's a discipline. One that starts with boundaries, intention, and, yes, even a little selfishness (the healthy kind).

Productivity Isn't About Being Busy

First, let's smash the idea that "being busy" is the same thing as being productive. It's not. In fact, busyness can be a trap. When you're leading remotely, it's easy to fall into a cycle of constant activity: replying to messages, jumping into meetings, checking dashboards, then wondering why nothing strategic got done by the end of the day.

Real productivity as a remote leader is about focus – not presence. It's about carving out space to think, plan, and lead instead of reacting to every ping like Pavlov's dog.

Here's how to shift into high-value mode:

- **Block time for deep work.** Treat strategy time like a sacred meeting. Protect it.

- **Batch your comms.** Set windows for checking emails and messages instead of living in your inbox.

- **Delegate ruthlessly.** If you're doing work your team can own, you're robbing them of growth – and yourself of time.

You don't need to be available 24/7 to be effective. You need to be intentional with your time.

Build a Sustainable Routine (That Actually Works)

One of the perks of remote work is the flexibility to design your day. One of the pitfalls? Flexibility without structure turns into chaos pretty fast.

Leaders need routines – not rigid schedules, but healthy rhythms that create flow and avoid decision fatigue. Try this framework:

- **Start-up ritual:** A consistent way to begin your day (stretch, plan, coffee, review goals).

- **Power blocks:** Schedule high-energy tasks during your most productive hours.

- **Shutdown routine:** Signal the end of your workday – even if your desk is in your bedroom.

And yes, wear real pants, at least sometimes. It helps trick your brain into work mode. (Plus, you never know when someone's going to turn their camera on.)

Say No More Often

Burnout thrives in the land of endless yeses. Yes to one more meeting. Yes to every request. Yes to being the last one online.

Here's the truth: your team doesn't need a leader who's always available – they need one who's present, focused, and energized. That means setting boundaries. That means saying no. That means closing Slack at the end of the day without guilt.

Healthy leaders model healthy behavior. When you respect your own limits, you give your team permission to do the same.

Watch for the Red Flags

Burnout doesn't show up all at once – it creeps in. One skipped lunch here, one extra-late night there, and suddenly you're Googling "symptoms of extreme exhaustion" between calendar invites.

Watch out for signs like:

- Constant fatigue, even after rest

- Feeling detached or cynical about work

- Struggling to focus or make decisions

- Dreading communication with your team

If you're hitting these signs, pause. Reevaluate. Talk to someone. Take a real break. Your productivity – and your leadership – depend on your well-being.

You Can't Pour From an Empty Coffee Cup

Here's the deal: you can have all the tools, strategies, and dashboards in the world, but if you're running on fumes, you can't lead effectively.

Being a remote project manager means you're not just managing tasks – you're modeling what healthy, effective remote work looks like. That means prioritizing your own energy, protecting your focus, and creating a sustainable pace for yourself and your team.

Remote leadership isn't about doing more; it's about doing what matters – consistently, sustainably, and with a little joy along the way.

Chapter 3: Setting Up Your Virtual Team for Success

You can have the best tools, the slickest dashboards, and a bulletproof Gantt chart – but if your team isn't set up for success, your remote project is already on shaky ground. Tools don't deliver projects – people do. And in the world of remote work, how you build, onboard, and organize your team matters more than ever.

Gone are the days of relying on proximity to build relationships or tossing new hires into the deep end with a quick tour of the break room and a "good luck." In a distributed setting, every step – hiring, onboarding, assigning responsibilities – has to be intentional. Otherwise, you risk ending up with a group of talented individuals operating in silos, all rowing in different directions (and one guy still trying to find the shared folder).

This chapter is all about getting it right from the start. Whether you're forming a brand-new team or reshaping an existing one for remote work, we'll look at how to lay the foundation for clarity, collaboration, and trust. From hiring the right people to creating onboarding experiences that actually prepare them for success, and from setting expectations to defining roles that leave no room for "I thought someone else was doing that," – we're covering it all.

Remote teams don't just become high-performing by accident; they're built that way – on purpose, with precision, and maybe just a little help from this chapter.

How to Build a High-Performing Remote Project Team

Let's be honest – remote teams don't magically gel just because you've added the right people to a Slack channel. In fact, without intentional structure and leadership, remote teams can feel more like loosely connected freelancers than a united force moving toward a shared goal.

So, what's the difference between a remote team that thrives and one that fizzles? Spoiler: it's not just talent. It's about alignment, trust, and clear purpose. A high-performing remote team isn't a happy accident – it's the result of deliberate team design.

Start With the Right Mix of Skills and Roles

A well-rounded remote team needs more than technical chops. Yes, you want experts in their craft – but you also want collaborators, communicators, and people who thrive without constant supervision.

When building your team, consider:

- **Hard skills:** the capabilities required to get the job done.

- **Soft skills:** the ability to communicate, adapt, and take initiative in a distributed environment.

- **Work styles and time zones:** think about how collaboration will happen across locations and hours. A team of brilliant people who can never meet is a coordination nightmare.

Diversity of experience, perspective, and personality also fuels innovation. Don't build a team of clones – build a team of complementary strengths.

Define the "Why" From Day One

People perform better when they know *why* their work matters. As a remote project manager, part of your job is connecting the dots between individual tasks and the bigger mission.

Before diving into the how and what, communicate:

- What the project aims to achieve

- Why it matters to the business or end users

- How each person contributes to that success

When people understand the "why," alignment becomes easier, and motivation becomes organic.

Establish Norms and Rituals Early

In co-located teams, culture often develops organically – Friday lunches, hallway chats, and recurring jokes. In remote teams, you have to manufacture those touchpoints more deliberately.

Consider implementing:

- **Kickoff rituals:** like a welcome call, a team charter, or a shared playlist

- **Working agreements:** e.g., how fast people should respond, what channels to use, how to handle blockers

- **Shared values:** even something as simple as "We show up on time and own our work" can unite your team

These norms provide a behavioral blueprint – and help turn a collection of individuals into a cohesive team.

Use Roles, Not Job Titles, to Clarify Ownership

High-performing teams don't waste time figuring out who's supposed to do what. Clear role definition is essential – especially when no one can peek over the cubicle wall to ask, "Is this you or me?"

Job titles are helpful but not enough. Get specific:

- Who's responsible for decisions?

- Who reviews and approves deliverables?

- Who communicates updates to stakeholders?

Use RACI models (Responsible, Accountable, Consulted, Informed) or simple role maps to avoid finger-pointing and fuzzy expectations later.

Hire for Initiative, Not Just Execution

In remote teams, the initiative is gold. You want people who raise flags before problems escalate, suggest improvements without being prompted, and don't wait to be told what to do.

During hiring or team formation, look for signs of self-direction. Ask questions like:

- "Tell me about a time you solved a problem without being asked."

- "How do you manage your workday when no one's watching?"

Execution is great, but ownership is better.

Hiring and Onboarding Remote Team Members Effectively

Hiring remotely isn't just a copy-paste version of in-office recruiting with a Zoom interview slapped on top. The stakes are higher, the signals are fuzzier, and the margin for error is slimmer. In a remote environment, a bad hire won't just sit awkwardly in the next cubicle – they'll quietly drift into disengagement, missed deadlines, and "Sorry, I didn't see that message" replies.

But here's the good news: when done right, remote hiring opens up access to incredible global talent, and remote onboarding – if thoughtfully structured – can set your team members up to succeed faster than traditional, office-bound methods. You just need the right strategy.

What Makes Remote Hiring Different?

Remote hiring goes beyond resumes and references. You're not just evaluating technical fit – you're assessing communication, autonomy, and time management in an environment without physical supervision.

According to a 2021 study by Harvard Business Review, the top predictors of success in remote roles are self-motivation, strong communication skills, and proactive problem-solving – even more so than experience or education. These qualities are hard to teach but easy to miss in a surface-level interview.

To identify them:

- **Structure your interview process to simulate remote collaboration.** Include async tasks like writing samples, recorded video introductions, or trial project briefs.

- **Ask behavior-based questions.** For example: "Tell me about a time you had to make a decision without input from your manager" or "How do you prioritize your work when everything feels urgent?"

- **Test for digital fluency.** If someone fumbles over basic tools like Google Docs or Slack, that's a red flag in a remote-first setup.

Remember: you're not hiring someone to fill a chair – you're hiring someone to fill a gap, take ownership, and collaborate confidently from Day One.

Onboarding: The Make-or-Break Phase

Onboarding isn't just a checklist – it's culture transfer. In remote teams, it's also the most common place where things quietly fall apart. Research from Gallup shows that only 12% of employees feel their company does a great job onboarding new team members – and that stat holds even truer for remote hires, who often face unclear expectations and isolation from the start.

Here's how to onboard remote team members effectively (and make them feel like part of the team before they've even touched a task):

1. Pre-Boarding Is Part of Onboarding

The experience starts the moment they accept the offer – not the moment they log in.

- Send a welcome email with the next steps, logins, and what to expect.

- Introduce them to the team in advance, even if just via a Slack channel or group message.

- Ship any hardware or equipment early (laptops, headsets, swag if you've got it).

Psychological research shows that people's first impressions significantly shape their long-term engagement. Starting strong matters.

2. Create a Clear, Repeatable Onboarding Plan

Don't wing it. Build a structured onboarding experience that spans at least the first two weeks, ideally 30-60 days.

Include:

- A written onboarding guide with key contacts, resources, and FAQs

- A detailed schedule of meetings, check-ins, and self-paced tasks

- Role-specific learning goals for the first 7, 30, and 60 days

Using platforms like Notion, Trello, or Google Sheets, you can build templates once and reuse them for every new hire – saving time while keeping quality consistent.

3. Assign a Buddy or Onboarding Partner

Onboarding is overwhelming. Assigning a peer – not a manager – can help new hires ask "small questions" without fear. This buddy can help them understand team norms, decode acronyms, and navigate your company's weird obsession with Monday meeting gifs.

According to research by Microsoft, employees onboarded with a peer buddy are 23% more satisfied with their onboarding experience and become productive faster.

4. Use a Blend of Synchronous and Asynchronous Learning

Remote onboarding works best when it mirrors how your team actually communicates. Use async video intros, recorded walkthroughs, or even short Loom tutorials to avoid calendar overload while still keeping things personal.

Pair that with scheduled check-ins:

- Day 1: Welcome and systems tour

- Day 3: Goals and role clarity

- End of week 1: First impressions and feedback loop

- Weekly for 30 days: Coaching, questions, and alignment

The goal? Prevent them from ever thinking, "Am I doing this right?" in silence.

5. Overcommunicate in Week One, Then Taper

In a remote setting, it's easy for new hires to feel like they're floating. Early overcommunication – "Here's what we're doing, here's why, here's how" – provides structure and confidence. But don't let that turn into micromanagement. As they gain comfort, they gradually shift toward autonomy.

Use a simple cadence like:

- Daily stand-ups (brief, async, or live)

- Mid-week syncs

- Weekly manager 1-on-1s

A great rule of thumb: when in doubt, communicate more – but always with purpose.

Bringing someone into a remote team isn't just about getting them logged in and assigned to a task. It's about helping them feel empowered, connected, and aligned from the beginning.

With intentional hiring and onboarding, you can avoid costly false starts, boost engagement, and create a team that hits the ground running – wherever they happen to be running from.

Establishing Clear Roles, Responsibilities, and Expectations

Remote teams thrive on clarity and crumble without it. When your team isn't in the same building (or even the same time zone), assumptions can become dangerous. "I thought someone else was doing that" is the remote equivalent of stepping on a LEGO barefoot: painful, frustrating, and completely avoidable.

In the absence of hallway conversations and spontaneous check-ins, you can't afford to let roles and expectations evolve "organically." You have to define them *intentionally* and document them clearly.

Why Clarity Matters More Remotely

In traditional teams, ambiguity can sometimes be patched over with quick chats and impromptu alignment. But in remote settings, vague expectations lead to missed deadlines, duplicated efforts, and team friction that builds in silence.

A study by Asana found that **26% of workers cite unclear responsibilities** as their top cause of missed deadlines. In remote teams, that percentage is often higher because the signals are harder to read.

Here's what clear roles and expectations accomplish:

- They reduce unnecessary questions and confusion.

- They eliminate duplicate work and dropped balls.

- They build a sense of ownership, autonomy, and accountability.

Define Roles Beyond Job Titles

Job titles are helpful, but they rarely tell the whole story. A "project coordinator" in one team might be managing timelines, while in another they're handling client comms. Avoid assumptions by spelling it out.

Start with a simple breakdown:

- **Role:** A one-sentence description of the person's key purpose on the team.

- **Primary responsibilities:** A list of what they're directly responsible for delivering.

- **Support responsibilities:** Where they play a secondary or collaborative role.

- **Decision rights:** What they can approve or decide without escalating.

Frameworks like **RACI** (Responsible, Accountable, Consulted, Informed) are especially useful in remote environments. They help teams understand:

- Who owns the task (Responsible)

- Who has the final say (Accountable)

- Who gives input (Consulted)

- Who needs to be kept in the loop (Informed)

Bonus: This also prevents the dreaded "too many cooks" problem.

Set Expectations Around Communication and Availability

In a co-located office, team norms emerge naturally. For example, everyone arrives around 9, lunch is at noon, and Steve always starts meetings with a joke that doesn't quite land.

Remote teams don't have that luxury. Instead, managers must create clarity around:

- Working hours and time zones

- Expected response times (e.g., "We reply to Slack within 24 hours")

43

- How to flag urgent issues

- Which tools to use for what

For example:

- Use Slack for quick questions and updates

- Use project management tools (like Asana, Trello, and ClickUp) for task-tracking

- Use email for formal comms and documentation

- Use video calls for brainstorming, decision-making, or anything nuanced

If expectations aren't defined, they'll default to chaos, and chaos in remote work is just silence with stress.

Co-Create the Rules (Don't Just Dictate Them)

Here's a secret: people are more likely to follow expectations they help create. As you define roles and responsibilities, involve your team in the process.

Try this:

- During project kickoff or onboarding, ask each team member to review and confirm their role description.

- Revisit roles when new projects begin, or team structures shift.

- Encourage feedback: "Does this responsibility list reflect what you're actually doing? What's missing? What's unnecessary?"

This collaborative approach builds buy-in, strengthens clarity, and ensures your systems reflect reality and not just what's on paper.

Document It All (Then Actually Use It)
Remote teams live and die by documentation. Once roles, responsibilities, and expectations are clear, write them down and make them easy to find. Not buried in an email thread or floating in a forgotten folder.

Use a shared team wiki, Notion, Google Docs, or any central platform your team already uses. Link to it in your onboarding guide, your project dashboards, and your team handbook.

Clarity isn't a one-time exercise. It's a system. The more transparent your team's structure, the faster people can make decisions, take ownership, and collaborate without constant hand-holding.

When everyone knows what they're doing, why it matters, and who's counting on them, something magical happens: the team starts leading itself. That's the sign of a high-performing remote setup, not when you're needed constantly, but when the system works smoothly, even without you.

Part 2: Managing Projects in a Remote World

Chapter 4: Remote Project Planning and Execution

Ah, planning – the part of project management where everything looks clean, logical, and achievable... right before reality kicks in. In remote teams, project planning takes on a new level of complexity: timelines get stretched across time zones, deliverables float between platforms, and stand-ups sometimes feel more like roll calls than productive touchpoints.

But here's the truth: remote work doesn't make project planning harder – it just makes poor planning more obvious.

The traditional project methodologies we've relied on for decades – Agile, Scrum, Waterfall – can still work beautifully. They just need a little remodeling to survive in the remote wild. In this chapter, we're going to take the best parts of these methods, strip away the assumptions of co-located teams, and rebuild a planning process that works *wherever your people are*.

You'll learn how to align your team around clear goals, define accountability structures that don't rely on hallway chats, and set up project kickoffs that energize (instead of confuse). Whether you're managing a sprint from Singapore or planning a product launch across five time zones, this chapter will help you bring clarity, structure, and flow to your remote execution.

Because a well-planned remote project isn't a unicorn – it's just a team with the right systems in place.

Let's get planning.

Adjusting Traditional Project Methodologies for Remote Teams

There's no one-size-fits-all approach to project management, and that's even more true in a remote world. The good news? Whether you're Agile to the core, Waterfall nostalgic, or somewhere in between, your favorite method can still work. It just needs to evolve with the way your team communicates, collaborates, and delivers.

Let's explore the three most common methodologies — Waterfall, Agile, and Scrum — and how to retool them for success with remote teams.

Waterfall: The Classic Blueprint (and How to Modernize It Remotely)

How It Works: Waterfall is a linear, sequential project methodology where each phase flows into the next like, well, a waterfall. You move through stages like requirements, design, implementation, testing, and delivery — each one typically completed before the next begins.

This approach thrives in environments where the scope is fixed, outcomes are predictable, and the process is king (think construction, manufacturing, or certain regulated industries).

Remote Adjustment Tips:

- **Break it into visible stages:** Use project management tools like ClickUp, Asana, or Wrike to make each phase visible to stakeholders and contributors. Add visual timelines to track handoffs and milestones.

- **Document everything clearly:** Waterfall depends on thorough documentation. In remote teams, make this documentation accessible, organized, and centralized – think shared folders, wikis, or Notion databases.

- **Increase checkpoints:** In an office, progress is often observed informally. Remotely, schedule regular check-ins to assess progress and surface issues early. These don't need to be meetings – status updates can be async.

- **Don't lock everything:** Even in a Waterfall project, allow for "controlled flex." Add buffer time between stages and build in short reviews for team alignment before each phase transitions.

Agile: Flexibility at Its Finest (With Remote-First Tweaks)

How It Works: Agile is all about iterative progress, collaboration, and responding to change over rigid plans. Agile teams work in short cycles (iterations), often delivering working versions of the product at each stage. It emphasizes individuals, customer collaboration, and adaptive planning.

It's perfect for environments with shifting priorities, fast feedback loops, and innovation-led projects (like software development, marketing campaigns, or product design).

Remote Adjustment Tips:

- **Lean into async tools:** Replace "hallway alignment" with structured stand-ups in tools like Slack (e.g., morning check-ins with what you did, what you're doing, and blockers).

- **Visualize the backlog:** Agile boards (Trello, Jira, Linear) become your virtual whiteboard. Make backlog items visible and prioritize ruthlessly.

- **Protect sprint boundaries:** When remote, it's easy for scope creep to sneak in. Hold the line – review scope changes in retrospectives, not in the middle of the sprint.

- **Create a feedback loop culture:** Use screen recordings, shared prototypes, or live demos to keep feedback flowing – without relying on back-to-back meetings.

Agile was born from the idea of adaptability. Remote work just asks you to adapt it even more – toward transparency, documentation, and fewer assumptions.

Scrum: Rituals With Remote Rigor

How It Works: Scrum is a popular Agile framework built around time-boxed sprints, defined team roles (Product Owner, Scrum Master, Dev Team), and recurring ceremonies: daily stand-ups, sprint planning, sprint reviews, and retrospectives.

It's great for fast-moving teams working toward a shared product vision, with clear deliverables and lots of iteration.

Remote Adjustment Tips:

- **Rethink your ceremonies:** Daily stand-ups can be async (e.g., posted in Slack or Teams). Keep sprint planning and retros as live calls if possible, but keep them focused and on time.

- **Shared dashboards = shared understanding:** Use Jira, Azure DevOps, or Shortcut to run virtual sprints. Everyone should know what's in the sprint and who's working on what – at all times.

- **Set a clear sprint rhythm:** Remote teams lose natural momentum without shared physical space. A predictable cadence (e.g., Monday sprint kickoff, Friday async demo) builds structure and team confidence.

- **Sprint retros go beyond metrics:** Use retros not just to optimize workflows but to check in on team health. Ask what's working – and how everyone's *really* doing. Bonus: tools like EasyRetro or FunRetro can help gather insights anonymously.

Choosing the Right Method (or Mix)

Don't be afraid to blend methodologies to suit your team. Many remote teams operate with a "Hybrid Agile" approach – borrowing Waterfall's structure for strategic planning, Agile's flexibility for development, and Scrum's rituals for cadence and accountability.

The key is intentionality. Whatever methodology you choose, define how it works *remotely* – and communicate that clearly.

Defining Clear Goals, Deliverables, and Accountability Structures

In a traditional office, it's easy to fake alignment. Everyone nods in meetings, scribbles notes, and walks away, assuming they're all working toward the same thing, until they aren't. In remote teams, you don't have that luxury. Misalignment isn't just frustrating – it's project kryptonite.

That's why defining crystal-clear goals, deliverables, and accountability structures isn't a "nice to have." It's the backbone of every successful remote project. When expectations are fuzzy, people drift. When they're clear, teams execute with focus – even from opposite sides of the planet.

Define Goals That Actually Mean Something

Goals aren't just a checkbox – they're the north star. But they only work when they're *clear, measurable, and visible* to everyone.

SMART goals (Specific, Measurable, Achievable, Relevant, Time-bound) are still the gold standard, especially remotely.

Example (Not SMART):

"Launch new customer onboarding flow."

Example (SMART):

"Design and deploy a new customer onboarding flow that reduces setup time by 30% for new users by July 31."

Now every team member knows:

- **What** needs to happen

- **Why** it matters

- **When** it needs to be done

- **How** success will be measured

Make goals visible in your project dashboard. Pin them in Slack. Reference them in retros. When goals are out of sight, they're out of mind – and so is alignment.

Clarify Deliverables (Don't Assume Anything)

"Deliverables" is a fancy way of saying: What will we actually hand off when this is done?

Remote teams can't afford to guess what "done" looks like. That's how you end up with beautiful work that solves the wrong problem – or worse, no work at all.

To define a deliverable clearly, include:

- **Format** (e.g., slide deck, video, code repo, user manual)

- **Owner** (who's delivering it)

- **Due date** (when it's expected)

- **Acceptance criteria** (how you'll know it's complete)

Example:

Deliverable: Finalized onboarding walkthrough video

Owner: Julia (Product Marketing)

Due: June 10

Acceptance Criteria: Includes an intro, three main product features, CTA, under 2 minutes, exported in .mp4 format, reviewed by the Product team

This removes ambiguity, sets expectations, and reduces the dreaded back-and-forth of "Can you tweak this just a little more…"

Build Accountability Without Hovering

In remote work, you don't need to *see* people working — you need to *see progress*. Accountability isn't about tracking hours; it's about creating a system where everyone knows what they're responsible for, and how their work connects to the team's success.

Best practices:

- **Assign ownership for every task or milestone** — not to "the team," but to a named person.

- **Track progress transparently** using project tools like ClickUp, Asana, or Jira. Everyone should know who owns what, what's in progress, and what's stuck.

- **Use regular check-ins (sync or async)** to discuss blockers, not just updates. The goal is progress, not status theater.

Example:

Instead of: "The design team is working on the website updates."

Try: "Carlos owns homepage redesign (due May 5), Mia owns mobile nav QA (due May 8), and all updates will be demoed during next Thursday's team review."

Bonus: Use Accountability Agreements during project kickoff, which are simple documents where each person confirms their role, responsibilities, and goals. This avoids confusion later and helps teams self-manage.

When goals are clear, deliverables are specific, and accountability is visible, you don't need to manage through reminders – you lead through results.

Best Practices for Virtual Project Kickoff Meetings

There's something magical about the beginning of a project. It's that hopeful moment when everyone's on the same page (or at least pretending to be), the vision is fresh, and the deadline still feels generous. In remote teams, though, the kickoff meeting isn't just a formality. It is an important foundational step. It sets the tone, builds momentum, and clarifies expectations before the work begins.

Unfortunately, too many virtual project kickoffs fall flat. They're either too vague, too long, too technical, or too awkward – especially when half the team has their cameras off and the other half is wondering what they're doing there.

Let's change that.

A great virtual kickoff can energize a team, align cross-functional partners, and launch your project with the clarity and confidence needed to succeed – even when you're spread across five time zones and three video platforms.

Here's how to make it happen.

1. Know the Purpose of Your Kickoff

First, let's define what a kickoff is *not*:

- It's not a status update.

- It's not a deep dive into every task.

- It's not a glorified calendar invite.

A kickoff is about alignment and connection. It's your chance to bring everyone together (virtually) to:

- Confirm the project's goals and desired outcomes

- Introduce team members and define roles

- Review scope, timeline, and major deliverables

- Establish communication norms

- Create a sense of shared purpose

Think of it like laying the tracks before the train leaves the station. Without it, things derail quickly.

2. Send a Pre-Kickoff Brief

Remote meetings work best when people show up prepared. Don't make your kickoff the first time anyone is hearing about the project.

A few days before the meeting, send out a short but clear pre-kickoff brief that includes:

- Project name and a one-sentence summary

- Why this project matters (business value or customer impact)

- High-level timeline or milestone overview

- Agenda for the kickoff meeting

- Who's attending and why

This gives everyone time to process the "what" so you can focus the meeting on the "how" and "who."

Pro Tip: Ask attendees to review the brief and come up with one question or potential risk they foresee. This gets people thinking ahead – and keeps them engaged during the call.

3. Curate Your Attendees Thoughtfully

Not everyone needs to be in the kickoff call. Include only those who are:

- Directly involved in the project

- Responsible for decision-making or deliverables

- Supporting the team in a strategic way (e.g., leadership sponsor)

Too many attendees = passive participants. A focused room (even a virtual one) is a productive one.

If you have a large stakeholder group, consider running a core team kickoff and a separate stakeholder alignment session.

4. Structure the Agenda Around Clarity and Connection

Every minute of your kickoff should build momentum and reduce confusion. A strong virtual kickoff agenda might include:

1. **Welcome & Purpose (5 mins)**

   Quick greeting, review of the agenda, and why this project matters.

2. **Team Introductions (5–10 mins)**

   Name, role, location, fun question ("What's your favorite productivity hack?"). Helps humanize the virtual room.

3. **Project Overview (10 mins)**

   - What are we building/solving?
   - What problem are we addressing?
   - What's the desired outcome?

4. **Scope & Deliverables (10 mins)**

   - What's included and what's not
   - Key deliverables and ownership

5. **Timeline & Milestones (10 mins)**

   - High-level phases
   - Key dates and dependencies
   - Known risks or blockers

6. **Roles & Responsibilities (10 mins)**

Review team structure and decision-making authority. Mention tools like RACI or a team charter.

7. **Communication Plan (5 mins)**

   - What tools will we use (Slack, Jira, Zoom, Notion, etc.)?

   - How we'll share updates

   - When we'll meet (cadence and format)

8. **Q&A and Open Discussion (10 mins)**

   Invite questions, risks, and suggestions.

9. **Next Steps (5 mins)**

   Confirm immediate actions, who's doing what, and when the next meeting is.

5. Set the Tone – Literally

People mirror the energy of the room – even on Zoom. So, if you're leading the kickoff, bring the vibe:

- Be enthusiastic, even if your team is still sipping their first coffee

- Show appreciation for people's time and talents

- Use humor when appropriate – it makes things memorable

- Encourage cameras on (but don't guilt people if they decline)

You're not just running a meeting; you're launching a shared journey. Set the emotional tone you want to carry through the project.

6. Use Visuals to Anchor Attention

In a remote setting, attention is your most valuable currency, and it's in short supply.

Share a simple slide deck or virtual whiteboard that outlines:

- The project's purpose

- Timeline

- Roles and contact points

- Major deliverables

You don't need to build a 40-slide presentation. In fact, please don't. Aim for 5-7 slides max that reinforce your talking points and keep everyone on track.

Tools like Miro, Figma, Google Slides, or FigJam can also help facilitate visual brainstorming or real-time alignment during the session.

7. Capture and Share the Kickoff Summary

The meeting shouldn't live and die in that Zoom room. After the kickoff:

- Send a summary within 24 hours

- Include key decisions, roles, goals, and next steps

- Link to relevant docs (e.g., project brief, team charter, RACI, backlog board)

- Record the meeting and share the link for those who couldn't attend

This creates a single source of truth for the start of the project and saves you from answering the same five questions three times next week.

8. Watch for Early Warning Signs
The kickoff is also a diagnostic tool. Pay attention to:

- Who's unclear about their role

- Who's not contributing

- Where tension, misalignment, or skepticism surfaces

These early signs will help you spot potential derailers and address them before they become real issues.

Keep a running list of questions and action items that pop up, and assign someone (maybe you, maybe not) to follow up on each one.

Virtual Kickoff = Culture in Action
A kickoff meeting is more than logistics – it's a signal. It tells your team:

- "We care about getting this right."

- "We're in this together."

- "We value clarity, collaboration, and purpose."

Do it well, and your project starts with energy, trust, and direction. Do it poorly, and you spend the next three weeks in damage control mode, untangling confusion that could've been avoided with 60 minutes of intention.

So bring the energy. Bring the structure. Bring the (virtual) donuts if you must.

Because a great kickoff isn't just the start of the project but rather, the foundation for how your team will work together until the final deliverable is shipped and celebrated.

Chapter 5: Communication Strategies for Remote Teams

If remote work had a greatest challenge, it wouldn't be time zones, technical issues, or finding a quiet space without a barking dog in the background. It would be communication.

In a traditional office, you can rely on casual chats, facial expressions, and overheard conversations to stay aligned. In a remote team? All you've got is what's typed, clicked, said on Zoom – or left unsaid. That makes communication both the glue and the grease of distributed collaboration. When it works, things run smoothly. When it breaks? Projects stall, people drift, and misunderstandings multiply faster than unread Slack messages.

Great communication doesn't just happen. It's not about adding more meetings or using more emojis (though we're not anti-emoji). Rather, it's about choosing the right tools, using them purposefully, and building a team culture where people share openly, ask questions early, and clarify constantly.

This chapter will help you cut through the noise – literally and figuratively. You'll learn how to choose the right channels for the right messages, when to go async versus live, and how to reduce the dreaded "I thought you meant…" moments that plague even the most well-meaning teams.

Because the best remote teams don't just talk a lot – they communicate with clarity, consistency, and intent.

Let's dive into how to make that happen.

Choosing the Right Tools: Slack, Microsoft Teams, Zoom, Email, and More

In remote teams, communication tools are your digital lifeline. They're the bridge between confusion and clarity, chaos and coordination. But with so many options – Slack, Microsoft Teams, Zoom, email, project management platforms, and more – it's easy to fall into the trap of tool overload. The key isn't to use every tool. It's to use the right tools in the right way for the right purpose.

This section breaks down the most common remote communication tools, their strengths, weaknesses, and best use cases – so you can build a streamlined communication stack that supports, rather than sabotages, your team.

Slack: The Digital Office Hallway

What it is: Slack is a real-time messaging platform built around channels for specific teams, projects, or topics. It's one of the most popular communication tools in remote work culture.

Best for: Quick team chats, casual communication, async check-ins, fast Q&As, link sharing, and integrations.

Benefits:

- **Real-time collaboration:** Instant messages feel like a virtual water cooler or hallway conversation.

- **Organized channels:** You can create dedicated channels for specific topics (e.g., #marketing, #project-x, #random) to reduce message scatter.

- **Powerful integrations:** Slack connects with tools like Google Drive, Jira, Asana, Trello, Zoom, GitHub, and more, making it a true communication hub.

- **Searchable archive:** Conversations are stored and searchable, helping teams avoid rehashing the same discussions.

Tips for using Slack well:

- Establish channel naming conventions (e.g., #proj-client-a, #ops-finance).

- Pin important messages so they don't get lost in the scroll.

- Use threads to keep discussions tidy.

- Set expectations around response time (e.g., Slack isn't 911).

Watch out for: Slack fatigue. Too many messages can create distractions and FOMO. Use do-not-disturb features and encourage async norms when appropriate.

Microsoft Teams: The Enterprise Communication Hub

What it is: Microsoft Teams combines chat, video conferencing, file sharing, and app integration under one roof – especially useful for organizations already using Microsoft 365.

Best for: Larger, enterprise-level teams that need robust security, file collaboration, and integration with Microsoft products.

Benefits:

- **Deep integration with Microsoft tools:** Word, Excel, PowerPoint, OneDrive, SharePoint – it's all there.

- **Structured team spaces:** Similar to Slack channels but tied to groups in your organization (e.g., by department or project).

- **Built-in video and voice calls:** You don't need to switch platforms for meetings.

- **Advanced security and compliance:** This is especially important for industries with strict data governance needs.

Tips for using Teams well:

- Leverage Teams tabs to add key documents, dashboards, or tools to your project channels.

- Use Planner integration for lightweight task tracking.

- Create dedicated private channels for sensitive discussions.

Watch out for: Clunky UX (especially early on), overlapping functions with other Microsoft tools, and occasional steep learning curve for new users.

Zoom: The Face-to-Face Stand-In

What it is: Zoom is the go-to platform for video conferencing, offering HD video, screen sharing, breakout rooms, and webinar capabilities.

Best for: Synchronous meetings, one-on-ones, team bonding sessions, workshops, presentations, and client calls.

Benefits:

- **Reliable video and audio quality:** Especially at scale.

- **Breakout rooms:** Great for workshops or small group collaboration.

- **Screen sharing with annotations:** Perfect for demos, walk-throughs, and brainstorming.

- **Recording and transcripts:** Let absent team members catch up later.

Tips for using Zoom well:

- Use the waiting room feature for security.

- Turn on live transcription for accessibility.

- Keep cameras on for small team meetings to boost engagement – but make it optional when people need space.

- Assign roles (host, co-host, notetaker) to keep meetings focused.

Watch out for: Zoom fatigue. Limit unnecessary meetings and batch sessions when possible, and encourage audio-only breaks.

Email: The Digital Paper Trail

What it is: Email remains the formal, long-form communication staple for most businesses.

Best for: External communication, formal updates, asynchronous announcements, documentation, and anything requiring a record.

Benefits:

- **Universally accessible:** Everyone has email.

- **Asynchronous by nature:** Reduces urgency and interruption.

- **Traceable:** Easy to forward, archive, and search for later.

- **Ideal for formal or legal documentation:** Contracts, approvals, summaries, etc.

Tips for using email well:

- Keep emails brief and clear – one main topic per message.

- Use subject lines like headlines.

- Use bullet points or numbered lists for readability.

- Clearly state any actions or deadlines.

Watch out for: Misuse as a default tool. Email isn't great for brainstorming, fast back-and-forths, or urgent communication.

Popular tools: Asana, Trello, ClickUp, Monday.com, Basecamp, Jira, Notion

What they are: These platforms manage tasks, timelines, deliverables, and responsibilities. Think of them as your remote command center.

Best for: Task assignment, progress tracking, project planning, dependencies, collaboration, and knowledge management.

Benefits:

- **Centralized visibility:** Everyone can see who's doing what and when.

- **Templates and workflows:** Standardize recurring processes.

- **Automation and reminders:** Reduce manual nudges.

- **Dashboards and reporting:** Offer instant project health updates.

Tips for using PM tools well:

- Define statuses clearly (e.g., To Do, In Progress, Blocked, Done).

- Limit boards to what's actionable – don't let them become cluttered archives.

- Tie all updates and decisions back to the tool (not just in Slack).

- Assign owners and due dates to everything.

Watch out for: Tool abandonment. If you pick a PM tool, live in it. Half-used boards and outdated tasks create confusion and mistrust.

Other Tools Worth Mentioning

Notion: All-in-one wiki, note-taking, task tracking, and document collaboration tool. Great for documentation, onboarding guides, SOPs, and team handbooks.

Loom: Async video recording tool perfect for walkthroughs, updates, and explainer videos. Reduces meetings and boosts clarity.

Miro / FigJam: Visual collaboration boards for brainstorming, diagramming, and mapping workflows.

Google Workspace: Real-time document collaboration. Docs, Sheets, Slides, and Drive make working together easy, especially for content or data-driven teams.

Calendly / SavvyCal: Async scheduling tools to reduce the back-and-forth of "When are you free?" – especially helpful across time zones.

Build a Communication Stack, Not a Frankenstein

It's tempting to add a new tool for every new problem – but the goal isn't more tools. It's fewer, smarter, integrated tools that serve your team's needs.

When choosing or refining your communication stack, ask:

- What do we need to communicate? (Status, feedback, updates, documents, deliverables, etc.)

- When does it need to happen? (Now, later, every week?)

- Who needs to know? (Individual, team, stakeholder?)

- What level of formality or documentation is required?

And don't forget to document your tool usage guidelines — what each tool is for, when to use it, and how to use it well. That clarity alone can save hours every week.

In summary, great communication in remote teams isn't about bombarding everyone with more messages but about choosing the right medium for the message:

- Slack or Teams for real-time collaboration

- Zoom for focused meetings and connection

- Email for formal or external updates

- Project management tools for task clarity and progress tracking

- Async tools like Loom and Notion for flexibility and documentation

When your communication stack is clear, consistent, and aligned with your team's needs, you don't just work better — you *feel* better.

Asynchronous vs. Synchronous Communication: When to Use Each

If remote work has a secret weapon, it's asynchronous (async) communication. And if it has a silent productivity killer, it's the misuse of synchronous (sync) communication. The trick is knowing when to use which – and how to design your workflows so they support clarity, not chaos.

Let's break it down, starting with a quick definition:

- **Synchronous communication** happens in real-time. Everyone is present (virtually or physically) and responding live – think Zoom calls, Slack huddles, or phone calls.

- **Asynchronous communication** happens with a time delay. One person sends a message or update, and the recipient responds when it works for them – think email, project comments, Loom videos, or task notes.

Both are essential. But using the wrong one at the wrong time? That's where projects (and patience) start to unravel.

The Case for Asynchronous Communication

Async communication is the backbone of high-functioning remote teams. It allows flexibility, encourages thoughtful responses, and eliminates the need to "be on" all the time. For distributed teams in multiple time zones, it's often the only practical option.

Benefits of async:

- **Respects deep work:** No constant pings or interruptions

- **Supports diverse schedules:** Ideal for global or flexible teams

- **Reduces pressure:** People can respond when they're ready, not mid-task

- **Encourages documentation:** Written updates can be reused and referenced

Best uses for async:

- Status updates and progress check-ins

- Feedback on documents or designs

- Questions that don't need an instant answer

- Project planning discussions

- Sharing recorded demos or walkthroughs

- Performance reviews (written summaries first)

Example:

Instead of a daily standup call, your team posts a short Slack update each morning:

- What I worked on yesterday

- What I'm working on today

- Any blockers

This takes 2 minutes, doesn't interrupt anyone's flow, and keeps the whole team in sync – on their own time.

While async is powerful, some things just need to happen live. Human connection, complex decisions, emotional nuance – these are better handled with real-time interaction.

Benefits of sync:

- **Faster decision-making:** Especially when the context is shared or urgency is high

- **Stronger relationships:** Face-to-face time (even virtually) builds trust and rapport

- **Nuance and emotion:** Tone, body language, and energy don't always translate in text

- **Great for brainstorming:** Collaborative thinking thrives in real-time

Best uses for sync:

- 1:1s and team bonding

- Sprint planning and retrospectives

- High-stakes or sensitive conversations

- Urgent decisions or blockers

- Brainstorming sessions or workshops

- Client or stakeholder presentations

Example:

You're launching a new product and need alignment from marketing, sales, and product. Instead of a week of email back-and-forths, you host a 45-minute Zoom to make decisions, clarify handoffs, and confirm the launch plan.

Real-time meetings still have a place – but the goal is intentional use, not default mode.

Choosing the Right Mode

When deciding between async and sync, consider the following:

1. **Urgency**

   - Can this wait a few hours? If yes, async.

   - Do we need a decision now? Go sync.

2. **Complexity**

   - Is it a simple update or question? Async.

   - Is it ambiguous, high-stakes, or emotional? Sync.

3. **Team availability**

   - Are we across multiple time zones? Async may be more inclusive.

   - Is everyone online and engaged? Sync can work.

4. **Documentation needs**

   - Should we have a record of this? Async or recorded sync.

   - Is it a one-time brainstorm? Sync may be fine.

5. **Cost of misalignment**

- Can miscommunication cause delays or errors? If yes, sync may be safer.

Building a Culture That Supports Both

The best remote teams don't just *default* to one mode — they design their culture to support both. That means:

- **Setting communication expectations:** Let people know what kinds of messages require immediate response and which don't.

- **Defining "core hours" (if needed):** A 2–4 hour overlap window for live collaboration across time zones.

- **Encouraging thoughtful async updates:** Celebrate clear, concise written updates the same way you praise a great meeting.

- **Being transparent about your schedule:** Use status tools (e.g., Slack statuses, calendar blocks) to show when you're heads-down vs. available.

- **Making it okay to decline a meeting:** Especially if it could've been an email or a Loom.

Pro Tip: Don't just assume people know when to use what. Document your team's communication preferences in your onboarding guide or team handbook.

Examples in Practice
Async-First Culture:

A globally distributed design team uses Loom for feedback, Figma comments for revisions, and a shared Trello board for task tracking. They hold one sync meeting a week, but most communication happens on their own schedule – leading to fewer meetings and more deep work.

Hybrid Culture:

A product team works across the U.S. and Europe. They use Notion to document specs, GitHub for dev collaboration, and Slack for questions. Each Wednesday, they sync for sprint planning. The rest of the week is async-first, with optional co-working hours.

Sync-Leaning Culture:

A customer success team with mostly overlapping hours uses Microsoft Teams to chat, call, and host daily huddles. Because they're client-facing, rapid responses and live collaboration are critical – but they still record important meetings and use email for formal follow-ups.

Common Pitfalls to Avoid

- **Using sync for everything:** This leads to burnout, meeting overload, and Zoom fatigue.

- **Going async with no structure:** Leads to silence, confusion, and dropped balls.

- **Expecting instant responses in async channels:** Undermines the purpose of async work.

- **Assuming everyone prefers your style:** Some team members thrive in async; others need regular check-ins to stay engaged.

Great remote teams don't communicate more – they communicate smarter. That means choosing between async and sync with intention.

Async gives you flexibility, space, and documentation. Sync gives you speed, connection, and clarity. Both are tools – *not* default settings.

When your team knows when to jump on a call and when to write a message instead, you reclaim hours, reduce stress, and empower everyone to work at their best pace – not just the fastest one.

Avoiding Miscommunication and Fostering Transparency

While good communication makes remote work work, miscommunication brings everything screeching to a halt. One misunderstood message, one vague comment, one missing update – and suddenly, your smooth-running project turns into a tangle of delays, frustration, and finger-pointing.

In an office, you might overhear someone struggling and jump in. Online? You often don't realize there's a problem until it's too late. That's why remote project managers must be proactive – not just in communicating but in building a system of transparency, clarity, and continuous alignment.

Here's how to prevent costly misunderstandings and create a team culture where information flows easily, and everyone stays on the same page.

Prioritize Clarity in Every Message

Vague messages are kryptonite in remote teams. Clarity isn't just polite – it's productive.

Tips for clearer communication:

- **Lead with the main point.** Don't bury the ask. Start with what you need, then add context.

- **Be specific.** Avoid words like "ASAP," "soon," or "touch base." Try "by Friday at noon" instead.

- **Use formatting:** Bullet points, headers, and bold text help highlight action items or deadlines in longer messages.

Example – Poor:

"Let's finalize the deck and get it to marketing quickly."

Example – Clear:

"Please finalize slides 1–10 and upload the final deck to Drive by Thursday at 3 PM. Once uploaded, tag the marketing team in Slack for review."

Bonus Tip: Use a team-wide writing tool like Hemingway Editor or Grammarly to help craft concise, readable messages.

Templates are your best friend when it comes to repeatable communication. They create consistency, save time, and remove ambiguity.

Useful templates for remote teams:

- **Project Brief Template** (Notion, Google Docs, or Confluence)

  - Project name and purpose

  - Stakeholders

  - Timeline and milestones

  - Success criteria

  - Risks and dependencies

- **Weekly Update Template**

  - What we did last week

  - What we're doing this week

  - What's blocked and why

  - Key decisions made

- **Meeting Notes Template**

  - Date/time, attendees

  - Agenda

  - Decisions

  - Action items (owner + due date)

- Next steps

- **Feedback Template**

  - What went well

  - What could be improved

  - Suggested next action

Keep all templates in a central location (Notion, Confluence, or Google Drive) and link them in your team's onboarding materials.

Make Statuses and Progress Visible

Transparency isn't about micromanaging but rather about reducing the need for constant check-ins.

Project visibility tools:

- **ClickUp, Trello, or Asana:** For visual Kanban-style task boards with clear ownership and due dates

- **Notion:** For tracking OKRs, timelines, and linked documentation in one place

- **Monday.com or Jira:** For complex dependencies or technical workflows

Visibility best practices:

- Tag owners and assign due dates for all tasks

- Use status updates like "In Progress," "Blocked," "Needs Review," and "Done."

- Require weekly async updates with what's been done, what's next, and what's stuck

Tip: Use a shared dashboard that is visible to all team members and stakeholders. This builds trust and reduces the number of "Hey, what's the status of…" messages.

Establish a "Default to Documentation" Culture
If it isn't written down, it doesn't exist.

Remote teams must make documentation part of the workflow – not an afterthought. This doesn't mean pages of paperwork. It means short, purposeful summaries of decisions, changes, and key knowledge.

What to document consistently:

- Meeting outcomes and action items

- Major decisions and the reasoning behind them

- Process changes

- Onboarding materials and role expectations

Where to document:

- Notion, Confluence, or Google Docs for knowledge hubs

- Slack threads (with pinned or saved highlights)

- GitHub wikis or README files (for dev teams)

Pro Tip: Assign a rotating notetaker for each meeting. Make summaries a routine part of your project process.

Create Communication Norms (and Write Them Down)
Don't assume everyone knows how your team communicates. Write it down in a Team Communication Charter and revisit it regularly.

What to include:

- What each tool is used for (e.g., "Slack is for questions and quick updates, ClickUp is for task progress")

- Response time expectations

- Preferred formats for updates (e.g., weekly roundup in Notion)

- Meeting etiquette (camera policy, agenda expectations, recording rules)

- Time zone norms and core working hours

Example entry:

"Async standups are posted in #standup by 10 AM local time, M-F. Use the template: Yesterday / Today / Blockers. If you're out, mark OOO in Slack status and calendar."

Post your charter in your onboarding guide and link to it often.

Foster a Feedback Culture

Miscommunication often festers when people are afraid to speak up. Encourage a culture where feedback – on projects, communication, and team dynamics – is welcomed, safe, and actionable.

How to foster feedback:

- Ask regularly: "What's unclear right now?"

- Include a "What could we improve?" section in retros and reviews

- Normalize correcting misunderstandings without blame

- Use tools like Officevibe, Polly, or Google Forms for anonymous check-ins

Sample feedback questions:

- "What's one thing we could communicate better?"

- "Is anything in our tools or updates confusing or redundant?"

- "What's one thing we should document more clearly?"

Lean on Visual and Async Tools for Clarity

Sometimes, words just don't cut it. Visuals can reduce misinterpretation by 90%.

Visual tools that help clarify:

- **Loom:** Record quick walk-throughs of tasks, dashboards, or bugs

- **Miro / FigJam:** Build visual workflows or brainstorm with your team

- **Lucidchart or Whimsical:** For architecture, logic flows, or org charts

Tip: If you're making a complex change or asking for feedback, show rather than tell. A 3-minute Loom recording often beats a 10-paragraph message.

Make transparency a habit, not a hero moment. Integrate it into your team's weekly rhythm:

Example weekly flow:

- **Monday:** Post weekly goals in Slack or Notion

- **Midweek:** Async updates (project boards, status pings)

- **Friday:** Quick win roundup or team check-in (written or video)

Other rituals that promote transparency:

- **Sprint demos:** Showcase what's been done, even if it's messy

- **Weekly wrap-up posts:** Summary of progress, priorities, and shout-outs

- **Leadership updates:** Monthly email or video recaps from leadership to keep everyone informed

Consistency reduces confusion. When people know where to look for answers, they stop guessing – and start executing.

Remote miscommunication is rarely about bad intentions – it's about missing context, assumptions, and vague instructions. But with the right systems, tools, and habits, you can build a remote team where transparency is the norm and clarity is contagious.

Investing in templates, documentation, communication norms, and visibility isn't busywork — it's the infrastructure that holds your team together across distances.

Chapter 6: Building Trust and Team Culture from a Distance

Ah, trust, the invisible thread that holds any team together, yet one of the trickiest things to weave when you're all scattered across time zones, home offices, coffee shops, or coworking spaces. This chapter discusses the art and science of building a thriving remote team culture and nurturing trust even when you can't share coffee breaks or quick chats by the water cooler.

Without the glue of face-to-face interaction, remote leaders have to work smarter to keep teams motivated, engaged, and emotionally connected. It's easy for virtual teams to slip into mechanical task execution, where everyone is just checking boxes without feeling part of something bigger. But here's the good news: trust and culture are not about physical proximity so much as they are about intentional actions, clear communication, shared values, and meaningful connections.

In this chapter, we'll explore how to encourage engagement and motivation when you don't share an office, how to choose virtual team-building activities that don't make people groan or roll their eyes, and how to navigate the delicate terrain of remote conflicts and tough conversations without missteps. You'll get practical strategies, ideas, and tools that will help you strengthen the bonds within your team, because even though you might be working from a thousand miles apart, your culture can (and should) feel tightly knit.

Encouraging Engagement and Motivation Remotely

Keeping a remote team engaged and motivated is like tending a garden you can't see every day. You can't rely on passing smiles, informal check-ins, or in-office perks; instead, you need to be intentional, creative, and structured in how you nurture enthusiasm and commitment. Without this care, it's easy for team members to drift into disengagement, lose sight of purpose, or feel disconnected from the larger mission.

Understand What Drives Each Person

Motivation isn't one-size-fits-all. Take time to understand what drives each team member: some thrive on public recognition, others on opportunities for skill development, and some simply want to know how their work fits into the bigger picture. A simple Motivation Profile Template can help:

Motivation Profile Template:

- Name:

- Top strengths:

- What excites me at work:

- What drains me at work:

- Preferred recognition style (public, private, written, verbal):

- Personal/professional goals:

Ask team members to fill this out and revisit it regularly. It gives you a roadmap to engage people on their terms.

Use Recognition and Celebration

When working remotely, you can't just drop by someone's desk to say "great job." That means you need formal and informal recognition systems. Use tools like Bonusly (peer-to-peer recognition), HeyTaco (fun Slack-based shoutouts), or even a shared "Wins and Shoutouts" Slack channel.

Practical tip: End your weekly team meetings by going around the (virtual) room and having each person recognize one teammate's contribution. Small but consistent recognition builds a positive atmosphere.

Make Work Visible

Engagement often dips when people feel their work disappears into a void. Use tools like Trello, Asana, or Jira to make project progress transparent. Set up visible dashboards that highlight achievements, progress toward milestones, and upcoming priorities.

Example: Create a public team board where each project has a card, and celebrate moving cards to "Done." Visually seeing progress is motivating.

Provide Growth and Learning Opportunities

One major driver of engagement is the opportunity to learn and grow. Offer small but regular learning opportunities: monthly "lunch and learn" sessions, sharing articles or podcasts, or even rotating leadership of team meetings to give people a chance to practice facilitation.

Template idea: **Quarterly Development Plan**

- Skills I want to build:

- Stretch projects/opportunities:

- Resources or support I need:

- Check-in date:

This keeps growth intentional and shows you care about their long-term development, not just task completion.

Use Pulse Surveys to Listen Actively

A key engagement driver is feeling heard. Tools like Officevibe, TinyPulse, or even a monthly anonymous Google Form can capture how the team is feeling. Ask about workload, stress, engagement, and suggestions for improvement.

Example survey questions:

- On a scale of 1 to 10, how engaged do you feel at work?

- What's been your biggest challenge this month?

- Is there anything we should change or improve as a team?

Act on the feedback quickly since people may disengage if they share concerns and nothing changes.

Align Work to Purpose

Remote workers are more engaged when they understand why their work matters. During team meetings or one-on-ones, connect daily tasks to the company's mission or big-picture goals. Share customer success stories, show the impact of their contributions, and remind them how their work moves the needle.

Practical tip: Start meetings with a "mission moment" – a short story, testimonial, or update that ties work back to purpose.

Encourage Social Connection

Motivation isn't just about tasks – it's also about belonging. Build small moments for personal connection:

- Create "watercooler" Slack channels for hobbies, pets, or casual chat.

- Host monthly virtual coffee chats or random "donut" pairings (using tools like Donut for Slack) where teammates meet informally.

- Celebrate birthdays, anniversaries, or milestones, even if it's just a fun emoji parade or a 5-minute shoutout.

Set Clear Expectations

Ambiguity kills motivation. Remote workers need clarity on what's expected, what success looks like, and how they're measured. Use Role Clarity Templates:

- Key responsibilities:

- Success metrics:

- Communication norms:

- Decision-making authority:

Review this with each team member and check in regularly to adjust as needed.

Provide Flexibility with Boundaries

One advantage of remote work is flexibility, but that only boosts engagement if people know it's okay to use it. Encourage flexible schedules, but set boundaries so no one feels pressured to be "always on." For example:

- Define "core hours" for collaboration, but allow flex time.

- Model behavior by not sending emails late at night.

- Use tools like Clockwise or Google Calendar Working Hours to protect focus time.

Lead by Example

Finally, remember that engagement is contagious. Show up with energy, authenticity, and commitment. Be transparent about challenges and wins, celebrate progress, and show your team that you genuinely care about their experience.

By using these practical tools and templates, you set up a framework where engagement doesn't happen by accident – it's deliberately designed into how you lead. Remote teams can be just as motivated, connected, and passionate as co-located teams when leaders take the time to design engagement thoughtfully. Let's keep building!

Virtual Team-Building Activities That Actually Work

Let's be honest, many remote workers hear "virtual team building" and immediately think of awkward icebreakers or cringeworthy forced fun. But here's the truth: when done well, virtual team-building activities can build trust, strengthen relationships, and inject some much-needed fun into a distributed team's routine.

So, what actually works?

Virtual Coffee Chats or "Donut" Pairings

One of the biggest challenges remote teams face is the lack of spontaneous conversations that happen in an office – hallway chats, lunchroom banter, or quick desk-side check-ins. To replicate this, tools like Donut (a Slack integration) randomly pair team members for short, informal coffee chats. These 15-30-minute chats help team members connect on a human level, especially across departments or roles that don't naturally interact.

Template Idea: Set up a "coffee chat calendar invite" template that auto-fills with prompts like "Share one fun fact" or "Talk about a recent challenge you overcame." Keep it light; no work talk is required.

Virtual Game Sessions (That Aren't Cheesy)

Games can lower barriers and build camaraderie – but only if they're chosen wisely. Trivia contests, online escape rooms, and collaborative games like Skribbl.io or Jackbox Games can create genuine laughter and teamwork moments. Schedule these occasionally (not too often!) and keep participation voluntary.

Practical Tip: Assign a rotating "Game Master" role so different team members get to choose or host the activity. This gives ownership and ensures variety.

Shared Learning Experiences

Instead of always focusing on social games, try bonding over learning. Host a monthly "Lunch & Learn" where a team member presents something – it could be a work topic, hobby, or even a book review. Alternatively, join an online workshop together (e.g., design thinking, leadership, or creative writing) and discuss takeaways afterward.

Template Idea: Create a simple sign-up form for Lunch & Learn slots and a feedback form to gather post-event thoughts.

Recognition Rituals

Building a culture of appreciation is essential, especially when people are isolated. Set up weekly or monthly shoutouts – using tools like Bonusly or simply a dedicated Slack channel – where team members recognize each other's contributions. You can also introduce fun awards like "Most Creative Solution" or "Best Virtual Background."

Practical Tool: Use Trello or a shared Google Doc to track recognition moments and ensure everyone is included over time.

Virtual Wellness Challenges

Fostering wellness can be a great bonding tool. Consider step count challenges, hydration reminders, or meditation streaks where the whole team participates. Tools like Strava, Headspace for Teams, or even a simple shared leaderboard can keep people engaged.

Example Challenge: "Walk 10,000 steps every day for a week and post a photo from your daily walk in the team chat. Winner gets a $20 coffee gift card."

Storytelling Sessions

Host a "Story Slam" or "Two Truths and a Lie" session where team members share surprising or funny stories. You can theme it around past work adventures, travel mishaps, or childhood memories. Storytelling builds empathy, understanding, and often some good-natured laughter.

Template Idea: Prepare a list of optional prompts ahead of time to help less talkative team members join in confidently.

Cross-Team Projects

Encourage people from different teams or departments to collaborate on a non-critical project, like redesigning the internal newsletter, creating a fun company playlist, or planning a virtual retreat. These give people a break from their regular roles while fostering creativity and bonding.

Practical Tip: Provide a clear brief, timeline, and deliverables – make it lighthearted, but treat it with enough structure to give purpose.

Effective virtual team-building isn't about overloading people with extra meetings, but about intentional, meaningful activities that strengthen trust and connection. Focus on activities that respect people's time, appeal to different personalities, and rotate leadership roles to keep things fresh.

If you're unsure where to start, survey your team: what kinds of activities excite them? What do they dread? Let their input guide your choices.

Remember, strong remote teams don't just happen – they're built, one shared laugh, challenge, and celebration at a time.

Handling Conflicts and Difficult Conversations Remotely

No matter how strong your team culture or how well you communicate, conflicts are inevitable, and when you're managing a remote team, handling them becomes even trickier. Without face-to-face cues, hallway chats, or the ability to pull someone aside for a quiet talk, misunderstandings can fester, tensions can rise, and small issues can quickly balloon into major problems. That's why remote leaders must be proactive, prepared, and skilled at handling difficult conversations across screens.

Recognizing Early Warning Signs

In a remote setting, conflicts often simmer beneath the surface before they erupt. Watch for signs like passive-aggressive messages, sudden changes in responsiveness, team members withdrawing from conversations, or subtle shifts in tone over email or chat. Create a culture where people feel comfortable voicing concerns early, and regularly check in with individuals to sense if something is brewing.

Practical Tip: Use a simple pulse survey (Google Forms, Typeform, or tools like Officevibe) every couple of weeks to gauge team sentiment. Include questions like, *"Do you* *feel any unresolved tensions within the team?"* or *"Is there anything making collaboration difficult right now?"*

Preparing for Difficult Conversations

When you need to address an issue, preparation is key. Set aside time to gather facts, review relevant communications, and clarify what outcome you want. Avoid making assumptions about someone's intent; instead, focus on observable behaviors and their impact.

Template Idea: Use the SBI (Situation-Behavior-Impact) framework to structure your feedback:

- **Situation:** Describe when and where the issue occurred.

- **Behavior:** Describe the specific behavior you observed.

- **Impact:** Explain the effect the behavior had on you, the team, or the project.

For example: *"In yesterday's planning meeting, when you spoke over Alex several times (Situation and Behavior), it caused frustration and made it harder for the team to hear all viewpoints (Impact)."*

Creating a Safe and Focused Space

Remote conversations need extra intentionality. Schedule a dedicated video or voice call (avoid handling conflicts over chat or email). Set a respectful, calm tone, and make sure both parties know the purpose of the conversation ahead of time. Ensure you're in a private, distraction-free environment, and encourage the other person to do the same.

Tool Tip: Use platforms like Zoom, Microsoft Teams, or Google Meet with breakout rooms or private meeting links to avoid drop-ins or interruptions.

Listening Actively and Navigating Emotions

One of the hardest parts of remote conflict resolution is reading emotional cues without physical presence. Practice active listening: repeat back what you hear, acknowledge feelings, and ask clarifying questions. Watch for voice tone, pauses, or visible frustration on video and address it gently.

Practical Idea: Try using a shared document (Google Docs, Notion) where both parties can note key points or agreements during the conversation. This helps keep the discussion grounded and creates a record of next steps.

Following Up and Documenting Agreements

A single conversation rarely resolves everything. After the discussion, send a brief summary email outlining what was discussed, agreements reached, and next steps. This provides clarity and accountability for both sides. Check in again after a week or two to ensure progress and keep lines of communication open.

Template Example:

Hi [Name],

Thanks again for taking the time to discuss [topic] today. Here's a quick summary of what we agreed on:

- [Point 1]

- [Point 2]

- [Next steps and timeline]

Let's check in again on [date] to see how things are going. Please feel free to reach out anytime before then if you have concerns or need support.

Best,

[Your Name]

Building a Long-Term Conflict-Resilient Culture

Beyond handling individual conflicts, aim to create a team culture that's resilient to disagreements. This means:

- Establishing clear team norms around respectful communication.

- Encouraging healthy debates and diverse viewpoints.

- Providing regular training or workshops on constructive feedback and conflict resolution.

- Modeling vulnerability and openness as a leader.

Tool Tip: Consider using platforms like Lattice, CultureAmp, or TinyPulse to gather ongoing team feedback, or run regular retrospectives (especially in Agile teams) to surface issues early.

Remember, the goal isn't to avoid conflict, but to handle it effectively. Managed well, even difficult conversations can strengthen relationships, improve processes, and increase trust. As a remote leader, mastering this skill will make you a more resilient and effective manager – and help your team thrive, no matter the distance.

Part 3: Overcoming Remote Work Challenges

Chapter 7: Managing Productivity and Performance

When it comes to managing remote teams, few topics stir up as much anxiety for leaders as productivity and performance. After all, how can you be sure people are working when you can't see them? Are they putting in the hours, or binge-watching TV with their Slack status set to "online"? These worries are common and understandable. But here's the truth: presence does not equal productivity, and micromanagement doesn't magically make people perform better. In fact, remote work challenges many of the traditional assumptions managers hold about performance, forcing us to rethink how we set expectations, measure outcomes, and keep teams focused across time zones and digital channels.

In this chapter, we dive into the practical side of boosting productivity and managing performance in a distributed environment. We'll cover how to set SMART goals (Specific, Measurable, Achievable, Relevant, Time-bound) that keep everyone aligned without drowning them in busywork. We'll explore strategies for evaluating performance based on results and not hours logged or keystrokes tracked, and we'll look at how to create a culture of trust where accountability flows naturally. Most importantly, we'll address the real-world headaches remote teams face: distractions at home, the blurring of work-life boundaries, and the logistical puzzle of coordinating people across multiple time zones.

Whether you're managing a fully distributed team or a hybrid one, this chapter will provide actionable tips, proven techniques, and ready-to-use templates to help you navigate these challenges. Our goal isn't just to make your team work harder but to help them work smarter, stay motivated, and deliver consistently high performance without burning out or losing sight of what really matters.

Setting SMART Goals for Remote Teams

Remote work demands clarity like never before. Without the physical cues and spontaneous check-ins of an office environment, team members rely heavily on defined direction to guide their efforts. That's why setting SMART goals — Specific, Measurable, Achievable, Relevant, and Time-bound — isn't just a productivity hack for remote teams, but an essential management strategy.

Why SMART Goals Work So Well Remotely

In a remote setting, misalignment can quietly derail progress. SMART goals help counter this by creating clear targets that keep everyone rowing in the same direction. They:

- Provide a shared understanding of what success looks like

- Create built-in accountability without micromanagement

- Help individuals prioritize and focus their efforts

- Give managers objective data for performance evaluations

Remote teams function best when they know what they're aiming for – and how to know when they've hit the mark.

Breaking Down SMART Goals

Let's take a closer look at each element of SMART goal setting:

S – Specific

A goal should be clear and unambiguous. Everyone should understand what is expected and why it matters.

Example:

- Not SMART: "Improve communication."
- SMART: "Implement a weekly asynchronous team update post in Slack every Friday by 3 PM to keep all team members aligned."

M – Measurable

There must be a way to track progress and determine when the goal is achieved.

Example:

- Not SMART: "Make the website better."
- SMART: "Redesign the homepage to reduce bounce rate by 20% within the next 6 weeks."

A – Achievable

Goals should be realistic based on available resources, time, and team bandwidth.

Example:

- Overreaching: "Launch three new product lines next month."
- SMART: "Complete market research and prototype one new product line for executive review by the end of the quarter."

R – Relevant

The goal should align with broader company objectives or team priorities.

Example:

- Misaligned: "Gain 5,000 new Instagram followers when your company sells B2B software."
- SMART: "Publish two LinkedIn thought-leadership posts per week for three months to drive inbound leads from target industries."

T – Time-bound

Every goal should have a deadline. Open-ended goals lack urgency and focus.

Example:

- Not SMART: "Update our help center."
- SMART: "Revise and publish 25 key help center articles by July 1, focusing on top 3 customer-reported pain points."

Setting SMART goals remotely involves intentional planning and regular communication. Here's a simple step-by-step framework:

1. Start with Team and Project Objectives

Before diving into individual goals, clarify the project or team-wide objectives. What are you collectively trying to achieve? This helps ensure alignment from the top down.

Example:

Objective: Increase customer satisfaction in Q3.

2. Involve the Team in Goal Setting

Collaboratively creating goals builds ownership and motivation. Instead of assigning goals from the top, co-create them during planning meetings or 1-on-1s.

Example:

"Based on our goal to improve CSAT scores, what specific actions do you think would make the biggest impact this quarter?"

3. Use Goal Templates and Trackers

Leverage a consistent goal-setting template so that everyone's goals are formatted the same way.

SMART Goal Template:

- Goal: [Full SMART statement here]

- Owner: [Team Member Name]

- Metrics for success: [How will we measure it?]

- Timeline: [Start and due dates]

- Related project/team objective: [Tie-in to broader goals]

Tools for tracking SMART goals:

- Notion or Confluence (for shared documentation)

- ClickUp or Asana (for tying goals to specific tasks)

- 15Five or Lattice (for tracking OKRs and performance reviews)

4. Review and Revisit Regularly
Remote teams need frequent alignment touchpoints. Set up regular goal reviews (weekly check-ins, bi-weekly retros, or monthly updates).

Practical Tip: Use a shared tracker or Kanban board where each person updates their goal status: Not Started / In Progress / At Risk / Complete.

SMART Goals in Action: Three Remote Role Examples
Software Developer

Goal: Complete refactoring of the user authentication module to improve page load speed by 40% by July 15.

- Measurable: Page load time benchmarked via Lighthouse reports

- Relevant: Supports platform performance and user retention

Marketing Manager

Goal: Launch a three-part email campaign targeting trial users, aiming to convert at least 15% to paid plans within 30 days of launch.

- Measurable: Conversion rate from trial to paid

- Time-bound: Campaign launches June 1, ends June 30

Customer Success Rep

Goal: Resolve 90% of support tickets within 24 hours for the top three issue types by the end of Q2.

- Specific: Focuses on top issue categories

- Achievable: Based on current ticket volume and staffing

Common Mistakes to Avoid

- **Setting vague goals:** "Improve the blog" means different things to different people.

- **Skipping timelines:** Without deadlines, even exciting goals lose momentum.

- **Confusing tasks with goals:** "Write three articles" is a task. "Increase organic traffic by 20% through new content" is a goal.

- **Overloading team members:** SMART doesn't mean stacking five ambitious goals at once. Focus on 1–3 high-impact goals per cycle.

SMART goals are more than a framework – they're a communication tool. They ensure everyone knows what's expected, how success will be measured, and why the work matters. For remote teams, that level of clarity is priceless.

By building a habit of setting and reviewing SMART goals, you'll foster alignment, boost motivation, and create a culture of accountability, all without hovering or guessing.

Ready to build performance without micromanaging? That's up next.

Measuring Performance Without Micromanaging

One of the biggest fears managers have when leading remote teams is losing visibility. Without the physical presence of an office, how do you know if people are working? How do you ensure that productivity stays high? And how do you balance oversight without creeping into micromanagement?

You can measure performance effectively without becoming a digital helicopter manager. In fact, the best remote leaders build trust by focusing on outcomes, not activity. They rely on clear expectations, transparent systems, and regular (but respectful) check-ins. In this section, we'll break down how to measure what matters, use the right tools, and foster accountability without undermining autonomy.

Why Micromanagement Doesn't Work (Especially Remotely)

Micromanagement is a fast track to disengagement. It signals a lack of trust, slows down progress, and increases stress. In remote teams, it's especially damaging because it leads to constant interruptions, over-monitoring, and decision bottlenecks.

Signs of micromanagement in remote settings:

- Asking for constant updates throughout the day

- Rewriting your team's work unnecessarily

- Insisting on being cc'd or looped in on every message

- Using surveillance tools to track screen time or mouse movement

Instead, great managers shift the focus from "Are they working right now?" to "Are we achieving the results we agreed upon?"

Shift from Time-Based to Outcome-Based Management

In remote environments, productivity should be measured by impact, not hours.

Ask yourself:

- Did the work move the project forward?

- Did the team meet the quality expectations?

- Did we hit our deliverables on time?

Example: Instead of tracking how long a designer is online, measure whether they delivered three high-quality mockups by Friday as planned, and whether the work aligns with the project's goals.

This shift not only empowers your team but also frees you to focus on coaching and strategic leadership instead of playing project traffic cop.

Set Clear Performance Metrics

To measure performance fairly and accurately, define clear success metrics for every role. These should be:

- Aligned to team and business goals

- Relevant to the role's core responsibilities

- Trackable using available data

Examples of remote-friendly performance metrics by role:

Developer:

- Code quality and review feedback

- Number of story points completed per sprint

- Uptime or performance improvements

Customer Support Rep:

- Number of tickets resolved

- CSAT (Customer Satisfaction) score

- First-response and resolution time

Marketing Specialist:

- Campaign ROI

- Engagement rate or click-through rate (CTR)

- Leads generated from organic content

Project Manager:

- Milestones delivered on time

- Budget adherence

- Stakeholder satisfaction (internal survey or feedback)

Pro Tip: Use SMART goals to create these metrics and tie them to quarterly objectives. This creates consistency across the organization.

Tools for Transparent Performance Tracking

Remote performance measurement is powered by systems, not surveillance.

Here are some tools that support visibility without being intrusive:

- **ClickUp / Asana / Trello:** Project management dashboards where progress is visible by task, assignee, and deadline

- **15Five / Lattice / CultureAmp:** Tools for weekly check-ins, performance tracking, and feedback

- **Notion / Confluence:** Centralized wikis or dashboards for goal tracking and milestone reporting

- **Google Sheets or Airtable:** Custom trackers for KPIs, individual goals, or team objectives

Dashboard Template for Managers (Notion or Sheets):

- Team member

- Goal / Metric

- Status (On Track / At Risk / Blocked / Completed)

- Last update date

- Next check-in

Having shared dashboards reduces the need for constant follow-up because everyone can see the same data.

Create a Regular Feedback Rhythm

Performance is best managed in conversations, not just scorecards. Set a rhythm for ongoing check-ins:

Weekly or bi-weekly 1-on-1s:

- Discuss priorities, blockers, and recent wins
- Ask open-ended questions like "What's been challenging this week?" or "What do you need more of from me?"

Monthly or quarterly performance reviews:

- Review goal progress and adjust as needed
- Celebrate achievements and growth
- Discuss long-term development goals

Quarterly retro templates:

- What went well?
- What didn't?
- What should we stop/start/continue doing?

Practical Tip: Use a shared 1-on-1 doc where both the manager and the team member can add topics before each meeting. This keeps the conversation productive and balanced.

Recognize and Reward Results

Don't let great performance disappear into the void. Acknowledge it!

Recognition ideas:

- Public shout-outs in team meetings or Slack channels

- Bonuses or rewards for hitting stretch goals

- Private notes of appreciation (especially meaningful from leadership)

Tools like Bonusly, HeyTaco, or a simple "#wins" channel in Slack help create a culture where results are celebrated, not just expected.

Support Underperformance with Coaching, Not Control

Not every team member will thrive immediately. If someone is underperforming, resist the urge to micromanage. Instead:

1. **Get curious, not critical.** Ask: "What's getting in your way?" or "How can I help you succeed?"

2. **Revisit expectations.** Were they clear, realistic, and documented?

3. **Create a support plan.** Include shorter check-in cycles, additional resources, or training.

4. **Track improvement.** Set a 30-60-90 day performance improvement plan (PIP) if needed.

PIP Template (simplified):

- Area of concern

- Expectations/goals

- Support offered

- Review dates

- Success criteria

The goal isn't to punish but to help the employee re-engage and contribute meaningfully.

Encourage Self-Assessment and Ownership

Performance management works best when team members feel ownership of their own progress. Encourage self-reflection by asking questions like:

- What are you proud of this quarter?

- What do you want to improve?

- What would you like to do more of or less of?

Self-Review Template (quarterly):

- Wins and accomplishments

- Challenges faced

- Lessons learned

- Goals for the next quarter

When employees participate in evaluating their own performance, it reduces defensiveness and deepens engagement.

Managing performance remotely isn't about watching people more closely – it's about empowering them with the right structure, clear expectations, and trust. When you focus on outcomes over activity, build systems for transparency, and give regular, honest feedback, you create a team that's motivated, accountable, and resilient.

Let go of control. Build systems that support visibility and growth. And remember: performance management should feel like coaching, not policing.

Overcoming Distractions and Time Zone Differences

Distractions and time zone differences are two of the most persistent hurdles in remote work. They silently chip away at productivity, collaboration, and team morale, especially if left unmanaged. The good news? With the right habits, tools, and systems in place, these challenges can be transformed from ongoing headaches into manageable nuances.

Let's break each one down and explore how to tackle them with intention.

Tackling Distractions: The Home Office Productivity Killers

Remote workers often face a different flavor of distraction than office-bound teams. Pets, children, household chores, noisy neighbors, and the siren call of the fridge or Netflix – these are all part of the remote reality.

Here's how to help your team minimize distractions and stay focused:

1. Help People Design Their Ideal Work Environment
Encourage team members to create a dedicated, distraction-free workspace, even if it's just a corner of a shared room.

Practical tips to share:

- Use noise-cancelling headphones

- Try ambient sound tools (like Noisli or Brain.fm)

- Use a room divider or visual cue to signal "focus time."

- Use a separate browser profile for work

2. Promote Time-Blocking and Focus Sessions
Time-blocking is one of the simplest and most effective strategies for remote productivity.

Example: Use Google Calendar or Clockwise to block 90-minute deep work sessions with labels like "No Meetings – Deep Work." Encourage your team to protect this time from interruptions.

Bonus: Use shared calendars to signal when you're in focus mode and not available for messages.

3. Introduce the Pomodoro Technique or Flow Apps
Pomodoro is a proven method for breaking up tasks into focused sprints (usually 25 minutes on, 5 minutes off). Tools like Pomofocus or Focus Booster can support this technique.

Apps like Serene or Forest gamify focus and discourage multitasking.

Distributed teams often span multiple time zones, and while this gives access to global talent, it can also create collaboration chaos if not handled thoughtfully.

Here's how to make time zones work for you, not against you:

1. Map the Time Zone Landscape
Start by identifying overlap hours, which are the windows of time where multiple team members are online.

Tool Tip: Tools like World Time Buddy, Timezone.io, or Google Calendar's "Working Hours" feature can help you plan meetings and deadlines across zones.

Example: If your team spans New York, London, and Bangalore, your overlap may be 8–10 AM ET. This becomes your golden collaboration window.

2. Set and Document Core Hours
Establish 2–4 hours of overlapping work time (if feasible), where team members are available for meetings or real-time collaboration.

Team Charter Template Addition:

Core Hours: Monday to Friday, 9-11 AM PST

Outside of these hours, async communication is the default.

3. Use Async Communication to Reduce the Need for Meetings
Rely on asynchronous tools for updates, decisions, and brainstorming:

- Use Loom or Bubbles for recorded video walkthroughs

- Post stand-ups and check-ins in Slack or Notion

- Use collaborative docs for project planning and comments

This allows team members to contribute in their own time without staying up at odd hours.

4. Plan Ahead for Handoffs

When team members are in vastly different zones, use a "follow-the-sun" model to keep work moving across time boundaries.

Practical Tip: Build handoff checklists or shared status boards that detail:

- What was completed

- What needs review or continuation

- Who's responsible next

Tool Suggestion: Trello, ClickUp, or Jira can be used to manage handoffs visually with clear ownership tags and comments.

5. Respect Boundaries and Avoid Burnout

Make it clear that no one is expected to work outside their local hours unless pre-agreed. Normalize asynchronous response times and lead by example.

Template Message for Slack or Email:

"No need to respond outside your local hours. Replying tomorrow is perfectly fine."

Distractions and time zones aren't going away. But with the right systems, your team can minimize their impact and stay productive without the pressure to be constantly online or perfectly synced.

By helping your team build better focus habits and design workflows around time zone realities, you unlock the true potential of remote work: flexibility, autonomy, and continuous progress, all without burning out or chasing people around the clock.

Chapter 8: Handling Remote Project Risks and Issues

Remote projects may come with freedom and flexibility, but they also come with a unique set of risks that can sneak up on even the most seasoned project managers. When your team is spread across cities, countries, and time zones – and your work depends on stable internet, good communication, and a little bit of trust – problems can escalate quickly if you're not ready for them.

This chapter is all about risk awareness, resilience, and readiness. You'll learn how to spot and manage the issues that are specific to distributed work: missed messages, delayed responses, flaky tech, time zone confusion, low visibility, and disengaged team members who quietly drift off the radar. These aren't just minor annoyances but can actually derail your project if you don't have the right systems and habits in place.

We'll start by exploring how to identify risks unique to remote projects, so you can be proactive rather than reactive. Then we'll dig into contingency planning; what to do when communication breaks down, a platform crashes, or half your team suddenly disappears into a digital black hole. And finally, we'll talk about how to lead with resilience when things go wrong (because let's face it, they will). You'll get practical ideas, real-world strategies, and templates to help you respond with calm instead of chaos.

Because in remote work, it's not about avoiding every hiccup, but trying to anticipate the turbulence, staying steady when it hits, and guiding your team through it with confidence.

Identifying Risks Unique to Remote Projects

Every project has risks, but remote projects come with their own special brand of unpredictability. While traditional risks like budget overruns, scope creep, and missed deadlines still apply, managing a distributed team introduces an entirely new layer of vulnerability. These risks aren't always loud or obvious. Often, they're subtle, cumulative, and silently corrosive if not addressed early.

As a remote project manager, your first line of defense is identifying these risks before they become real issues. Let's take a look at the most common ones that plague remote teams and how to spot them before they do damage.

1. Communication Gaps and Message Misfires

What it is: Important messages get buried, tone is misinterpreted, or information is delayed due to asynchronous schedules or a lack of clarity.

Why it's risky: Decisions are stalled, work is duplicated, or worse, team members operate with incorrect assumptions, leading to rework and conflict.

Early warning signs:

- "I didn't see that message."

- "I thought someone else was handling it."

- Long delays between messages and actions

Risk reduction:

- Use centralized communication tools (e.g., Slack, Teams) with channel-specific discussions

- Establish communication protocols and expectations for response times

- Use written recaps and action summaries after every meeting

2. Tech Dependencies and Platform Failures

What it is: Over-reliance on a specific platform (Zoom, Google Drive, Trello, etc.) that crashes, lags, or becomes inaccessible.

Why it's risky: Work halts, meetings are missed, or data is temporarily (or permanently) lost.

Early warning signs:

- "I can't access the file."

- "The meeting link isn't working."

- Frequent syncing or login issues

Risk reduction:

- Always have backup tools/platforms for core activities (e.g., Zoom + Google Meet)

- Store files redundantly (e.g., local + cloud copies)

- Create a tech contingency plan outlining what to do if a platform goes down

3. Team Disengagement or Isolation

What it is: Team members start checking out – emotionally or behaviorally – due to lack of interaction, purpose, or visibility.

Why it's risky: Lower performance, lack of accountability, and eventual turnover

Early warning signs:

- Less participation in meetings or asynchronous channels
- Missing deadlines without explanation
- Lack of enthusiasm or initiative

Risk reduction:

- Schedule regular check-ins (both 1:1 and group)
- Use tools like Officevibe or TinyPulse to track team sentiment
- Celebrate wins publicly and often
- Watch for drops in communication and follow up with empathy, not suspicion

4. Time Zone Silos and Collaboration Delays

What it is: Work gets bottlenecked because team members can't collaborate in real-time, or decisions are delayed due to scheduling gaps.

Why it's risky: Missed handoffs, duplicated work, misalignment

Early warning signs:

- Work is "waiting on feedback"

- Unclear handoffs between team members in different zones

- Missed deadlines due to misaligned priorities

Risk reduction:

- Map out your team's working hours and identify overlap windows

- Design workflows that don't depend on synchronous input

- Use async video tools (Loom, Bubbles) and shared task boards for handoffs

5. Lack of Visibility into Progress

What it is: Managers and stakeholders can't easily tell who's doing what, what's blocked, or how far along a deliverable is.

Why it's risky: Missed deadlines, lack of accountability, surprises at project reviews

Early warning signs:

- Constant status-check messages like "Any update on this?"

- Unclear ownership of tasks or priorities

- Stakeholders asking for progress reports last minute

Risk reduction:

- Use shared dashboards (ClickUp, Asana, Notion) with visible task owners and statuses

- Standardize weekly or bi-weekly status updates

- Implement project health reports with traffic-light indicators (Green/Yellow/Red)

6. Decision-Making Bottlenecks

What it is: Decisions are delayed because it's unclear who owns them, or approvals are stuck in email purgatory.

Why it's risky: Momentum stalls, scope expands, and team frustration builds

Early warning signs:

- "Who's signing off on this?"

- Deadlines slip due to feedback loops

- Endless comment threads with no resolution

Risk reduction:

- Use RACI charts to define decision roles (Responsible, Accountable, Consulted, Informed)

- Assign decision deadlines along with task deadlines

- Encourage empowered ownership with clear approval boundaries

Remote work doesn't introduce *more* risk than in-office projects. It just introduces different types of risk that require proactive visibility, intentional systems, and trust-based communication.

As a remote project manager, your superpower is your ability to anticipate these issues before they disrupt your momentum. Build a habit of scanning your project environment regularly for these subtle risk indicators, and you'll be better equipped to lead your team through challenges with calm, clarity, and confidence.

Contingency Planning for Communication Breakdowns, Tech Failures, and Disengagement

In remote project management, the unexpected isn't a question of *if*, it's a question of *when*. From dropped video calls and unsent Slack messages to missing team members and a surprise Wi-Fi blackout during a critical review, problems will arise. But you don't have to panic when things go sideways – if you have a contingency plan.

Contingency planning is the process of preparing backup strategies and protocols so that when the unexpected happens, you already know what to do. In this section, we'll look at three common disruptions – communication breakdowns, tech failures, and team disengagement – and walk through how to prepare for each.

1. Communication Breakdowns

What Can Go Wrong:

- Slack or Teams outage

- Messages are lost, misinterpreted, or ignored

- Delayed responses during time-sensitive tasks

- Language barriers or tone misunderstandings

1. Multi-Channel Communication Plan

Set expectations for which tools to use when. For example:

- Slack or Teams for day-to-day discussion
- Email for formal communication and backup during outages
- Project tools (Asana, ClickUp) for tracking decisions and tasks

Template Snippet for Team Handbook:

If Slack goes down, all team communications will move to email until service is restored. If video calls fail, default to Google Meet. Keep key project updates mirrored in project management software to maintain visibility.

2. Clear Communication Protocols

Establish guidelines for:

- Response time expectations (e.g., reply to urgent DMs within 2 business hours)
- How to escalate issues (e.g., tag the manager in the task comment or escalate via email)
- Meeting summaries and action items posted in a central location (Notion, Confluence)

3. Emergency Contact Tree

Create a simple contact matrix that outlines how to reach key personnel if standard tools are down. Include:

- Secondary emails

- Phone numbers (if appropriate)

- Backup team leads

2. Tech Failures

What Can Go Wrong:
- Internet outages

- Platform crashes (Zoom, Google Drive, etc.)

- Device or software malfunctions

Contingency Tactics:
1. Redundancy is Your Friend

Where possible, ensure the team has alternatives:

- Provide backup tools (e.g., both Zoom and Google Meet links in calendar invites)

- Maintain a local copy of essential project files in addition to cloud storage

- Use tools like Loom to pre-record walkthroughs in case you can't join a live session

2. Document Tool Failover Plans

Include fallback options in your tech stack documentation:

- Video conferencing: Zoom → Google Meet

- File sharing: Google Drive → Dropbox

- Messaging: Slack → Email

3. Maintain Offline Productivity Kits

Encourage team members to have a plan for working offline:

- Local copies of current tasks or documents

- Screenshots or PDFs of the current sprint boards

- Personal task lists in tools like Notion or Todoist

Bonus Tip: Include a "What to do if your internet goes out" checklist in your onboarding material.

3. Disengagement and Team Silence

What Can Go Wrong:
- A team member stops communicating

- Participation drops off

- A high performer's quality declines without explanation

Contingency Tactics:
1. Build in Early Warning Systems

Use lightweight tools like:

- Weekly async check-ins (e.g., via Range or Friday.app)

- Sentiment trackers (Officevibe, Lattice Pulse Surveys)

- Anonymous feedback forms

Encourage managers to watch for subtle signs: fewer messages, reduced meeting attendance, missed deadlines, or a drop in contribution quality.

2. Normalize Direct Check-Ins

If someone seems disengaged, check in privately and early. Use a soft opener like:

"I've noticed you've been a bit quiet lately. Is everything okay? Is there anything you need from me or the team?"

Keep it curious and supportive, not accusatory.

3. Document Escalation Paths

If disengagement persists, have a system for escalating gently:

- First, check in with the individual
- Follow up with the team lead or manager
- Discuss a support plan or temporary adjustment of responsibilities

Disengagement Follow-Up Template:

- Date of first concern
- Observable behavior changes
- Notes from 1:1 conversations
- Agreed next steps (shortened hours, added support, time off)

A well-prepared remote project team doesn't scramble in a crisis – they pivot. Having contingency plans in place keeps your team steady and solution-focused, even when your main tools go offline or when a key contributor disappears without warning.

By documenting your backup plans, building visibility into team wellbeing, and setting expectations up front, you empower your team to recover quickly and confidently from common disruptions. Think of these plans as your digital seatbelts: you hope you won't need them, but you'll be glad they're there when you do.

How to Maintain Resilience When Things Go Wrong

No matter how thorough your planning or robust your tools, things will go wrong in remote projects. A key hire might back out at the last minute. A product launch could flop. A week of miscommunication might snowball into a serious delay. In these moments, resilience is what sets successful remote leaders apart.

Resilience is your ability to stay steady, make decisions with clarity, and keep your team engaged even when the project hits turbulence. And in remote teams – where uncertainty and ambiguity can feel even more intense – it's one of the most essential leadership traits you can develop.

1. Normalize the Bumps in the Road

Start by acknowledging that setbacks are normal. Share this mindset with your team so that when challenges happen, they're not seen as personal failures, but part of the process.

Team message example:

"Hey everyone, our timeline's shifting, and that's okay. We're adjusting, learning, and continuing forward together. Let's stay focused on solutions."

This models calm and keeps people from spiraling into blame or panic.

2. Use Post-Mortems as Learning Tools
Instead of moving on quickly from failure, run short, structured retrospectives or post-mortems:

- What went wrong?
- What did we learn?
- What would we do differently next time?

Keep it blame-free. Focus on systems and improvements.

Pro Tip: Use templates like a "Start/Stop/Continue" or "5 Whys" framework to guide the conversation.

3. Communicate with Transparency and Empathy
In tough moments, silence is dangerous. Share what's happening honestly, but with reassurance and direction.

Crisis communication tip:

- Be factual about the situation
- Acknowledge impact
- Share next steps
- Offer support

Your tone sets the emotional bar. Calm, compassionate leadership keeps your team grounded.

4. Protect Energy and Morale

Don't expect everyone to push through nonstop. Build resilience by helping your team pace themselves:

- Encourage mental health breaks

- Use no-meeting days or deep work blocks

- Celebrate effort, not just wins

Sometimes, the best productivity boost is permission to pause.

Resilience isn't about being unshakable – it's about being able to bend and bounce back. When your team sees that you stay composed and solution-focused in challenging times, they learn to do the same.

Remote leadership is as much about emotional endurance as technical precision. Lead with calm, communicate with clarity, and turn every obstacle into an opportunity to grow stronger together.

Remote projects come with a unique set of risks, but they're not unmanageable. The key to success isn't eliminating every possible problem (which is impossible), but building the awareness, systems, and habits to respond quickly and effectively when challenges arise. From identifying early warning signs like disengagement or tech dependence, to creating practical contingency plans for communication breakdowns and platform failures, proactive preparation gives your team a safety net and you, as a leader, peace of mind.

When things inevitably go wrong – and they will – resilience becomes your most valuable asset. Staying calm, transparent, and focused in the face of setbacks helps your team stay motivated and solution-oriented, even during tough stretches. By creating a culture that learns from mistakes, plans for disruptions, and supports each other under pressure, you turn remote project risk into a strategic advantage, not a ticking time bomb.

Chapter 9: Running Effective Remote Meetings

In a remote work environment, meetings can either be the glue that holds your team together or the black hole where productivity disappears. We've all been there: another back-to-back video call with no clear purpose, attendees half-listening with their cameras off, and someone asking, "Wait, what did we decide?" That's not collaboration – that's burnout. The reality is, remote meetings require more structure, intentionality, and clarity than their in-person counterparts. When done well, they can drive alignment, foster connection, and keep projects moving. When done poorly, they lead to Zoom fatigue, frustration, and wasted time.

In this chapter, we'll tackle how to make your remote meetings not just tolerable, but *actually useful*. First, we'll explore how to eliminate the dreaded Zoom fatigue by designing meetings that respect your team's energy and time. You'll learn how to avoid unnecessary meetings altogether and replace them with asynchronous alternatives when possible. Then, we'll cover the foundations of effective remote meeting practices: choosing the right format, setting clear agendas, managing participation, and making space for both voices and focus.

Finally, we'll dive into the power of modern tools – AI assistants, recordings, automatic summaries, and transcriptions – that can enhance your meetings without adding extra work. With the right strategies and tools, you can turn meetings into one of your most effective remote management levers and not just a calendar filler.

Whether you're leading a daily stand-up, hosting a project kickoff, or facilitating a stakeholder review, this chapter will give you the templates, tools, and techniques to run meetings with purpose, impact, and efficiency. The goal isn't to have *more* meetings; it's to have better ones.

How to Eliminate Zoom Fatigue and Wasted Meeting Time

Let's face it, if remote work had a villain, it would be Zoom fatigue. Endless video calls with no breaks, no clear agenda, and no real outcomes drain energy faster than a Monday morning inbox. Unlike in-person meetings, virtual ones demand intense focus, fewer nonverbal cues, and more screen time – all without the casual social interactions that make in-office meetings bearable. The result? Exhaustion, disengagement, and a team that dreads every calendar invite.

So, how do you fix it? It starts with one radical idea: not every meeting needs to exist. Many "meetings" are just decisions that could've been made in a shared doc, a status update that fits neatly into Slack, or a question that didn't need a 30-minute slot. The first step in eliminating Zoom fatigue is ruthless prioritization. Before scheduling, ask: *Is this meeting necessary? Can it be async? Who actually needs to be there?* If you can't justify it, cancel it. Your team will thank you.

When a meeting is necessary, keep it focused and structured. Use a meeting agenda template with three sections: *Purpose*, *Discussion Topics*, and *Desired Outcomes*. Share it at least 24 hours in advance, and stick to it. Limit meetings to 25 or 50 minutes instead of the full half or full hour. That 5–10 minute buffer gives people time to reset and prevents the dreaded Zoom-marathon feeling.

Lastly, encourage camera-optional culture. Yes, face-to-face time matters – but not at the cost of comfort and concentration. Instead of requiring "cameras on," ask for "voices in." When people feel trusted and respected, they engage more, not less.

Zoom fatigue isn't inevitable. With a few intentional tweaks, your meetings can go from draining to dynamic and maybe even enjoyable.

Meeting Best Practices: When, How, and Why to Meet

In remote teams, meetings should be a tool, not a routine. A well-run meeting can align a team, clarify blockers, and drive progress forward. A poorly planned one? That's just 45 minutes of collective confusion. The stakes are higher in distributed environments because time zones, digital fatigue, and limited visibility mean that every meeting must pull its weight.

Here are 10 remote meeting best practices to ensure every session is intentional, impactful, and (dare we say) productive:

1. Always Know the "Why"

Before scheduling, ask yourself: *What is the purpose of this meeting?* If you can't clearly answer that, don't schedule it. Meetings without purpose become habitual calendar clutter. Is this to make a decision, share an update, brainstorm ideas, or strengthen relationships? The answer shapes the format.

2. Use an Agenda – Always

Never enter a meeting without a shared agenda. It doesn't have to be fancy. A simple doc or Notion page with a title, key points, time estimates, and expected outcomes is enough. Send it 24 hours in advance so attendees can come prepared.

Agenda Template:

- Objective of the meeting

- Items to discuss (with time blocks)

- Who's leading each item

- Action items & next steps

3. Invite Only the Necessary People

Smaller meetings are more focused. Only invite those directly involved in the decision or discussion. For larger updates, consider asynchronous recordings or follow-up summaries instead of pulling in the whole team.

4. Establish Ground Rules

Set basic norms like:

- Cameras are optional but appreciated

- Mute when not speaking

- No multitasking

- One conversation at a time

If you're running recurring meetings, create a shared etiquette doc for consistency.

5. Use the Right Meeting Type

Match the meeting type to the need:

- **Daily stand-ups** → Quick, sync check-ins (10–15 mins)

- **Weekly team meetings** → Recaps, planning, light collaboration

- **Project retros** → In-depth reflection, problem-solving

- **1:1s** → Personal support, feedback, career growth

- **Brainstorming sessions** → More interactive, whiteboard-friendly

Different purposes require different structures and tools – don't default to a "talking circle" every time.

6. Timebox Everything

Be ruthless with time. Assign limits to agenda items and stick to them. A 30-minute check-in should *not* become a 90-minute rabbit hole. Use a facilitator or timekeeper to keep things moving, and park off-topic discussions in a "side thread" to revisit later.

7. Start and End on Time

Punctuality shows respect for people's time. Start at the scheduled time – don't wait for stragglers – and wrap up with 5 minutes to spare for action items and questions.

8. End with Clear Outcomes

Don't let meetings fade out. Always end with:

- Key decisions made

- Action items (with owners and due dates)

- Next steps or follow-ups

Use a shared doc or task manager (like Asana, Notion, or ClickUp) to assign and track these items immediately.

9. Consider Async Alternatives

If a meeting's primary goal is to share information, ask: *Could this be a Loom video, a Slack update, or a Google Doc comment thread?* Async meetings are inclusive, time-zone friendly, and reduce overload.

Examples:

- Project status → Trello board + short Loom update

- Brainstorming → Miro board + async comments

- Feedback → Google Doc with comments instead of live edits

10. Review and Improve Regularly

Take 5 minutes once a month to review your meeting calendar:

- Which meetings are still useful?

- Which could be shorter or less frequent?

- Which can be replaced with async updates?

Ask your team for input – what's working, what's draining, and what would help them feel more engaged?

Remote meetings aren't just a replacement for office time – they're a chance to be *better* than office meetings. With clear purpose, lean invitations, and structured flow, your meetings can become moments of clarity and collaboration – not calendar clutter. The goal isn't to have perfect meetings. It's to have purposeful ones that move your team forward and respect their time.

Using AI, Recordings, and Summaries to Improve Meeting Efficiency

In remote work, meetings aren't just moments of conversation – they're digital assets. Every discussion, decision, and update carries value. But too often, those insights vanish as soon as the "Leave Meeting" button is clicked. That's where AI, recordings, and summaries come in. These tools don't just capture the conversation; they amplify it, reduce redundancy, and free up time for focused work.

Here's how to use them strategically to supercharge your meeting efficiency.

1. Recordings Are Your Asynchronous Superpower

One of the simplest ways to improve meeting effectiveness is to hit "record." Recording a meeting ensures that anyone who couldn't attend still has access to the full context. It also reduces the need for follow-up meetings and minimizes the "Can you fill me in?" messages that slow teams down.

Best Practices for Recordings:

- Record only when necessary (e.g., project kickoffs, stakeholder reviews, retrospectives).

- Always inform participants at the start.

142

- Store recordings in a central, organized repository (e.g., Google Drive, Notion, Confluence).

- Title recordings clearly: *[Date] – Project Update – Client A.*

Pro Tip: Encourage async feedback on recorded meetings. Team members can comment directly in the meeting doc or channel instead of needing a second meeting to weigh in.

2. Let AI Handle the Note-Taking

AI meeting assistants like **Otter.ai**, **Fireflies.ai**, **Grain**, and **Fathom** can automatically transcribe meetings, summarize key points, and even tag action items. This means you no longer need to assign a dedicated note-taker, and no one has to frantically type while trying to participate.

Benefits:

- Get searchable transcripts of every meeting.

- Auto-generated summaries save time on follow-ups.

- Action items can be pushed directly into project tools like Asana or Trello.

Use Case Example:

After a sprint planning meeting, Fireflies emails the team a summary with:

- Decisions made

- Tasks assigned (with owners)

- A link to the full transcript and recording

This eliminates confusion and ensures accountability, without extra admin work.

3. Use Summaries to Replace Recaps and Repeats

Instead of holding separate follow-up meetings, send a summary document or use a summary bot to distribute key takeaways. This works well for teams in multiple time zones or anyone catching up after PTO or illness.

Meeting Summary Template:

- Date & Meeting Title

- Attendees

- Discussion Summary

- Decisions Made

- Action Items (with owners and deadlines)

- Recording/Transcript Link

You can automate this in Notion, Google Docs, or tools like Fellow or Hugo. For recurring meetings, create a standard summary doc and update it in each session.

4. Keep a Knowledge Trail

One of the underrated benefits of using recordings and summaries is that you create a searchable archive of decisions. When someone new joins the team, or a stakeholder wants to know *why* something happened, you've got the receipts — without needing to rehash it from memory.

Tool Tip: Create a shared "Meeting Library" in Notion or Confluence where all key meetings are logged by project or topic. Tag them with keywords for easy access later.

5. Balance Efficiency With Human Connection

Just because you *can* automate doesn't mean you should always do it. Use AI to reduce administrative load, but don't let it replace real conversation. For emotionally sensitive topics or coaching moments, skip the transcript and stay present. The goal is to enhance communication, not depersonalize it.

Meetings don't have to be a time sink. With the help of smart tools – AI note-takers, automatic summaries, searchable recordings – you can create clarity, reduce repetition, and make every meeting more valuable. When you stop spending time capturing what happened and start focusing on what's next, your team works smarter, faster, and with more confidence.

Part 4: Advanced Strategies for Remote Project Success

Chapter 10: Leveraging Technology to Stay Ahead

If remote work is the new normal, then technology is its lifeline. The right tools don't just support distributed teams; they empower them. Whether it's automating repetitive tasks, streamlining project visibility, or protecting sensitive data, smart tech choices can make the difference between a remote team that's barely treading water and one that's setting the pace for innovation and growth.

In this chapter, we'll dive into how to leverage modern technology not just to manage projects but to stay ahead of the curve. We'll start by exploring AI and automation tools that can take some of the day-to-day burden off your shoulders, from summarizing meetings and assigning tasks to tracking project risks and forecasting timelines. These tools can free up time, reduce manual error, and let you focus on high-impact work.

Next, we'll take a closer look at the top project management platforms – Jira, Trello, Asana, ClickUp – and how to integrate them seamlessly into your workflow. It's not about using more tools... It's about using the right tools in the right way to ensure your team operates like a well-oiled machine, no matter where they are.

Finally, we'll cover a topic that's easy to overlook but absolutely critical in remote work: cybersecurity. From password hygiene to secure file sharing and access control, protecting your team's data is not optional. It's your responsibility.

By the end of this chapter, you'll have a practical tech playbook that helps you lead smarter, move faster, and stay secure in a constantly evolving digital landscape. Because in the remote world, the teams that adapt to technology don't just keep up – they lead the way.

AI and Automation Tools That Make Remote Project Management Easier

Managing projects remotely means juggling countless tasks, time zones, communications, and documents, all without the benefit of being in the same room. That's where artificial intelligence (AI) and automation come in. These technologies don't just "make things easier" – they act as digital team members, helping with everything from task management and note-taking to risk analysis and workload balancing.

In this section, we'll break down some of the most impactful AI and automation tools for remote project managers, exploring what each one does, the benefits it brings, and the potential trade-offs to consider.

1. ClickUp AI

What it does: ClickUp is a powerful project management platform that includes built-in AI tools for automating repetitive work, summarizing tasks and meeting notes, and providing real-time insights into project progress.

Benefits:

- Natural language processing to convert meeting summaries into action items

- AI-generated task updates and priority suggestions

- Automation of task creation, due date adjustments, and progress tracking

Disadvantages:

- Requires initial setup time and process mapping

- It can be overwhelming due to its many features

Best for: Teams already using ClickUp looking to level up with intelligent workflows

2. Fireflies.ai

What it does: Fireflies records, transcribes and summarizes voice conversations from Zoom, Google Meet, Microsoft Teams, and more. It also automatically detects action items and decisions.

Benefits:

- Real-time transcription and searchable meeting history

- Highlights tasks and follow-ups

- Integrates with Slack, Notion, and CRM platforms

Disadvantages:

- Accuracy may vary depending on speaker clarity

- Requires clear consent and privacy awareness in some jurisdictions

Best for: Project managers who attend frequent meetings and want automated, reliable records

3. Motion

What it does: Motion uses AI to automatically prioritize and schedule your tasks and meetings in real-time, based on your deadlines and focus hours.

Benefits:

- Automatically reorder your calendar as new priorities arise

- Helps reduce decision fatigue and planning overhead

- Can boost deep work time by optimizing your day

Disadvantages:

- Less effective for highly unpredictable schedules

- It may require manual override for non-flexible meetings, or client calls

Best for: Busy remote leaders who want calendar automation and focused productivity

4. Otter.ai

What it does: Otter transcribes meetings, interviews, and discussions in real-time and allows users to highlight, comment, and summarize content.

Benefits:

- Live transcription and collaborative note-taking

- Shareable notes with audio timestamps

- Works well for multilingual and international teams

Disadvantages:

- Requires a paid plan for full-feature access

- Quality depends on audio clarity and accents

Best for: Teams that need searchable, real-time meeting records and live collaboration

5. Loom AI (Async Video Messaging)

What it does: Loom enables quick async video messages for updates, reviews, and training – with AI-generated summaries, transcripts, and action points.

Benefits:

- Saves time and reduces the need for live meetings

- AI auto-generates transcripts and TL;DRs

- Enhances team communication with visual + verbal context

Disadvantages:

- It is not ideal for sensitive discussions or two-way interactions

- Still requires discipline to manage and organize shared videos

Best for: Explaining complex ideas or updates without needing a live call

6. Trello Automation (Butler)

What it does: Butler is Trello's built-in automation bot that uses natural language rules to automate task workflows.

Benefits:

- Automates recurring tasks (e.g., due date reminders, status changes)

- Sends rule-based notifications or updates to the team

- Simple to use for non-technical users

Disadvantages:

- Limited to Trello (no cross-platform automation)

- Automation logic can be rigid compared to custom tools

Best for: Teams using Trello who want to reduce manual task tracking and process upkeep

7. Notion AI

What it does: Notion's AI assistant helps generate summaries, write project plans, brainstorm content, and convert messy notes into structured documentation.

Benefits:

- Drafts clear project documentation or meeting recaps

- Extracts action items from notes

- Suggests improvements to unclear writing or planning docs

Disadvantages:

- Still evolving and sometimes generates generic outputs

- AI-generated content needs review and human editing

Best for: Knowledge-focused teams who already use Notion as a central hub

8. Zapier (Automation Integration Platform)

What it does: Zapier automates workflows by connecting thousands of apps together (e.g., Slack → Trello → Gmail). It creates triggers and actions across platforms.

Benefits:

- Automates handoffs between apps (e.g., new form → task created)

- Great for lead tracking, notifications, file backups, etc.

- No-code setup with templates

Disadvantages:

- It can become complex and hard to debug

- It may require paid plans for advanced features or higher usage limits

Best for: PMs who want to automate admin work between disconnected tools

9. Reclaim.ai

What it does: Reclaim uses AI to schedule and protect time for meetings, habits, and tasks based on your actual calendar and work goals.

Benefits:

- Automatically books time for focus, 1:1s, breaks, or lunch

- Rebalances schedule dynamically as priorities shift

- Helps remote workers maintain boundaries and avoid overload

Disadvantages:

- Integration is limited to Google Calendar (currently)

- It may require trust and calibration to avoid excessive rescheduling

Best for: Individual contributors and team leads who want to protect work-life balance and deep work time

10. Hypercontext

What it does: Hypercontext helps you run better 1:1 and team meetings by setting shared agendas, tracking goals, and logging feedback. AI assists in summarizing insights.

Benefits:

- Keeps recurring meetings consistent and focused

- Tracks team goals and progress from meeting to meeting

- Provides prompts and conversation starters for managers

Disadvantages:

- Best suited to leadership and middle management roles

- It may not fully replace project-tracking tools

Best for: Managers and team leads running multiple 1:1s or performance-based meetings

11. Jira Smart Automation & AI Features

What it does: Jira is widely used for software development and Agile project management. Its built-in automation engine allows you to set rules and workflows that auto-update tasks, assign work, and notify stakeholders based on triggers like issue status or custom fields.

Benefits:

- Automates issue transitions, assignments, and notifications

- Integrates with Git, Confluence, Slack, and CI/CD pipelines

- Customizable rules with no coding required

- Built-in AI (Jira Product Discovery) suggests insights from customer feedback.

Disadvantages:

- Steep learning curve for non-technical users

- It can be overly complex for simple projects

Best for: Software and product teams running sprints, tracking bugs, and managing engineering work at scale

12. Asana AI & Automation

What it does: Asana is a versatile work management platform with smart project tracking, rule-based automation, and built-in AI features for productivity and planning.

Benefits:

- Automates task handoffs, notifications, and updates

- Smart due date adjustments based on workload

- AI-generated project insights and risk alerts (in premium tiers)

- Clean interface with a robust template library

Disadvantages:

- Premium features are behind a paywall

- Some users may outgrow its simplicity for technical workflows

Best for: Cross-functional teams looking for a user-friendly platform that combines structure with flexibility

AI and automation aren't just buzzwords – they're productivity accelerators for remote teams. By offloading routine tasks, improving clarity, and providing smart insights, these tools free up your time for leadership, strategy, and human connection. Whether you're summarizing meetings, scheduling your week, or automating your admin work, these digital assistants can help your team run leaner, smarter, and more focused.

Integrating Project Management Platforms (Jira, Trello, Asana, ClickUp)

Using a project management platform is only half the battle; the real power lies in integration. Remote teams often juggle multiple tools: messaging apps, document hubs, calendars, CRMs, and developer tools. If your project management system doesn't talk to the rest of your tech stack, you end up with information silos, duplicated work, and a lot of copy-pasting. Integration closes that gap.

In this section, we'll explore how to get more out of Jira, Trello, Asana, and ClickUp by connecting them to the tools your team already uses and how to choose the right platform based on your needs.

Why Integration Matters for Remote Teams

Remote teams need clarity and flow. Integrating your project management platform allows:

- Seamless task creation from messages or emails

- Real-time updates across tools (e.g., Slack + Trello)

- Fewer context switches between apps

- Better data visibility for leadership and stakeholders

Jira Integration Highlights

Jira excels for technical teams, especially when integrated with:

- **Slack / Microsoft Teams**: Create and update issues directly from chat.

- **GitHub / Bitbucket**: Auto-link commits, branches, and pull requests to Jira tickets.

- **Confluence**: Embed Jira issues into documentation; use pages to track sprint progress.

- **Zapier**: Connect Jira to Google Sheets, Gmail, or even Trello for reporting or cross-functional syncing.

Pro Tip: Use Jira Automation rules to update task status when a branch is merged or a PR is approved.

Trello Integration Highlights

Trello is a lightweight visual platform that is great for marketing, admin, or early-stage project teams. Key integrations include:

- **Slack**: Turn messages into Trello cards with one click.

- **Google Drive / Dropbox**: Attach files directly to cards for reference.

- **Butler Automation**: Build automation within Trello (e.g., move a card to "Done" when a checklist is complete).

- **Zapier / Make.com**: Trigger workflows across email, calendars, or CRMs.

Pro Tip: Set up Trello to auto-update stakeholders by moving cards to a "Client Review" list that triggers an email or Slack ping.

Asana Integration Highlights

Asana strikes a balance between structure and flexibility. Its integrations support diverse teams:

- **Google Workspace**: Sync tasks with Google Calendar and attach Docs to tasks.

- **Slack / MS Teams**: Create or complete tasks from messages.

- **Notion / Loom**: Link project briefs or async video updates.

- **Harvest / Everhour**: Track time spent on tasks for billing or workload balancing.

Pro Tip: Set up Asana Rules to automatically assign tasks based on custom fields (e.g., priority or department).

ClickUp Integration Highlights

ClickUp is the most comprehensive tool on this list, with docs, goals, chat, and tasks all in one. Integrations enhance its capabilities:

- **Slack**: Turn messages into tasks or receive notifications for updates.

- **Google Calendar**: Sync due dates and create meeting-based action items.

- **GitHub / GitLab**: Track code deployments and pull requests directly in ClickUp.

- **Zapier / Make / Integrately**: Build complex multi-step workflows (e.g., form submission → ClickUp task → Slack alert).

Pro Tip: Use ClickUp's native email integration to send emails from within a task and log client communication.

Choosing and Connecting the Right Platform

Choose your tool based on:

- **Team type**: Jira for devs, Asana for cross-functional, ClickUp for all-in-one power, Trello for simplicity.

- **Existing stack**: Pick the platform that plays well with your calendar, file system, and communication channels.

- **Growth needs**: Consider scalability and permission settings as your team grows.

Start small: Integrate your core tools (e.g., calendar, chat, cloud storage), then expand to CRM, HR, or finance as needed.

The best project management tool isn't just the one with the most features but the one that connects your ecosystem. Integration reduces friction, increases visibility, and helps remote teams move as one. By setting up the right connections, you create a streamlined, automated workflow that minimizes manual updates and maximizes focus and collaboration.

Cybersecurity Best Practices for Remote Teams

When your team is distributed across homes, coworking spaces, cafes, and maybe even the occasional beachside Airbnb, cybersecurity becomes both more critical and more complicated. Unlike in a traditional office setup where IT manages secured networks, locked-down devices, and in-person access control, remote teams operate in a decentralized environment where every team member becomes a potential vulnerability point.

The risks are real: phishing scams, data leaks, weak passwords, compromised devices, and unsecured Wi-Fi connections can all lead to costly security breaches. For remote project managers, cybersecurity isn't just IT's job; it's a leadership responsibility.

Here's how to create a strong security culture that keeps your team's tools, data, and clients safe without killing productivity.

1. Start with a Cybersecurity Culture

Security is about more than the software your team uses. It's about habits. The first step is to foster awareness and ownership across the team.

Best Practices:

- Include basic cybersecurity training in onboarding.

- Talk about security openly and regularly (in all-hands, retros, etc.)

- Share stories or case studies of real-world breaches to show relevance

Tool Tip: Use platforms like Curricula, KnowBe4, or Wizer for gamified, remote-friendly security training.

2. Use Strong Passwords and Two-Factor Authentication (2FA)

One of the simplest and most effective safeguards is requiring strong passwords and enabling 2FA across all accounts.

What to do:

- Require team members to use a password manager like 1Password, Bitwarden, or LastPass

- Enforce 2FA on Google Workspace, Microsoft 365, project tools, and password vaults

- Prohibit password reuse (use vault-generated passwords)

Pro Tip: Avoid shared spreadsheets of passwords. Instead, use shared vaults for team logins with role-based access.

3. Secure Devices and Connections

Remote team members work from a variety of devices and networks, some secure and some sketchy. Protect endpoints and connections with a few simple steps.

Device Best Practices:

- Require full-disk encryption on laptops (standard on macOS, available on Windows)

- Use antivirus and anti-malware software (e.g., Malwarebytes, Norton, or CrowdStrike)

- Enable auto-lock after inactivity

- Keep all software and OS patches up to date

Network Best Practices:

- Avoid public Wi-Fi unless connected through a VPN (e.g., NordVPN, ExpressVPN, or a company-managed VPN)

- Disable auto-connect to known networks

4. Control Access to Tools and Data

Not everyone needs access to everything. Use role-based access control (RBAC) to assign permissions based on job function, not convenience.

Action Steps:

- Use SSO (single sign-on) where possible to centralize access management

- Set up tiered permissions in tools like Jira, Asana, ClickUp, and cloud storage platforms

- Immediately disable accounts when someone leaves the company

Tool Tip: Consider using admin panels and audit logs in platforms like Google Workspace or Microsoft 365 to monitor access and activity.

5. Use Secure File Sharing and Storage

File sharing is critical for remote teams but also a common security gap.

Best Practices:

- Use encrypted cloud services like Google Drive, OneDrive, or Dropbox Business.

- Avoid sending sensitive files via email – use expiring shared links instead.

- Limit who can view/edit/download files

- Enable version control to protect against accidental changes or deletion

Bonus Tip: Label files containing sensitive data and educate the team on how to handle them.

6. Prevent Phishing and Social Engineering

Phishing is the #1 cause of data breaches globally. Remote teams are especially vulnerable because attackers exploit distance, impersonation, and urgency.

How to Defend:

- Train your team to recognize suspicious emails, links, and requests

- Use email platforms with built-in threat detection (e.g., Google Workspace Advanced Protection)

- Simulate phishing tests to reinforce awareness

Red Flags to Watch For:

- Urgent requests for payment or password changes

- Emails that spoof internal users or leadership

- Unexpected attachments or shortened links

7. Create a Security Playbook for Remote Teams

If something goes wrong, your team should know exactly what to do. A simple incident response plan prevents panic and minimizes damage.

Include:

- Who to contact in a breach or suspected phishing

- What to do if a device is lost or stolen

- How to report suspicious activity

- Password reset procedures

Store this playbook in an easily accessible, secure location (e.g., Notion or Confluence with view-only permissions).

8. Audit Regularly and Update Policies

Security isn't one-and-done. Conduct periodic audits of your tools, devices, and access permissions.

Checklist for Quarterly Security Review:

- Are all users using 2FA?

- Are old accounts disabled?

- Are devices updated and encrypted?

- Has your team completed any security refreshers?

- Any unusual activity in admin logs?

Assign a security champion on your team or partner with your IT provider to run these reviews.

9. Balance Security with Usability

Too many security hoops can lead to "shadow IT," where employees use unauthorized tools to get around friction. Instead, make secure behavior easier by:

- Offering password managers instead of manual tracking

- Pre-configuring VPNs or SSO

- Using friendly tools with built-in protection (e.g., Slack over random chat apps)

Rule of Thumb: If a secure option is harder to use, adoption will suffer. Prioritize security and user experience.

Remote teams move fast, but cybersecurity must move faster. As a project manager, you don't have to be a tech expert to lead secure practices. What matters is creating a culture of awareness, using the right tools, and putting clear systems in place.

Cybersecurity should be treated as a mindset that protects your projects, your clients, and your team's reputation. With a few smart moves, you can turn your remote team's digital sprawl into a secure, resilient ecosystem.

Chapter 11: Leading Remote Agile & Hybrid Teams

Agile methodologies have long been praised for their ability to improve flexibility, increase collaboration, and deliver value faster. But what happens when your Agile team isn't gathered around a whiteboard in a shared office but spread across time zones, working asynchronously from their homes? Can stand-ups still stand up? Can sprints still sprint?

The good news: yes, Agile can thrive remotely if adapted thoughtfully. While traditional Agile practices were born in co-located teams with daily in-person collaboration, the core values – communication, feedback, adaptability, and continuous improvement – are just as powerful (if not more so) in remote environments. What needs to change is *how* those principles are executed.

In this chapter, we'll explore what it takes to lead high-performing Agile and hybrid teams in a distributed world. First, we'll look at how to adapt foundational Agile principles – like collaboration, iteration, and reflection – to remote contexts without losing their spirit. Then we'll dive into the day-to-day: how to run remote stand-ups, sprint planning, demos, and retrospectives in a way that's inclusive, efficient, and energizing.

Finally, we'll share real-world case studies of remote and hybrid teams successfully implementing Agile at scale. From startups to enterprise teams, you'll see how others are overcoming common challenges – like lack of visibility, slower feedback loops, and team cohesion – and what strategies you can borrow or build upon.

Whether you're managing a fully remote dev team, a hybrid product squad, or a distributed cross-functional unit, this chapter will give you the tools, tweaks, and templates to make Agile work *wherever* your team works because Agile isn't just a process. It's a mindset, and it's one that remote teams need now more than ever.

Adapting Agile Principles to Remote Work

Agile was born out of a need for flexibility and responsiveness in software development, but its principles apply just as powerfully across remote and hybrid teams in every industry. However, working remotely forces teams to rethink how Agile works in practice. When daily face-to-face interactions become, asynchronous check-ins and physical whiteboards are replaced by digital ones, the core values of Agile need to be translated for a virtual setting.

Let's take a closer look at the 12 principles of the Agile Manifesto and how each can be adapted for success in remote environments.

1. Customer satisfaction through early and continuous delivery of valuable software

Remote Adaptation:

- Use cloud-based tools to ship and deploy more frequently (e.g., GitHub Actions, CI/CD pipelines).

- Involve stakeholders asynchronously through recorded demos or shared dashboards.

- Replace live presentations with async Loom videos or annotated walkthroughs.

2. Welcome changing requirements, even late in development

Remote Adaptation:

- Create a shared change request process in tools like Jira, ClickUp, or Notion.

- Use collaborative backlog grooming sessions (live or async) to review evolving priorities.

- Keep user feedback loops open with in-app surveys, feedback forms, or customer support syncs.

3. Deliver working software frequently, with a preference for shorter timescales

Remote Adaptation:

- Automate your release and testing processes to support more frequent pushes.

- Use clearly defined sprint goals and smaller work units that can be reviewed quickly via screen recordings or live demos.

- Share delivery status via Slack bots or integrated status pages.

4. Business people and developers must work together daily throughout the project

Remote Adaptation:

- Use persistent team channels in Slack or Teams for daily communication.

- Pair stakeholders with developers using async tools like Loom or Google Docs for ongoing collaboration.

- Create virtual office hours or shared calendars to make stakeholders accessible.

5. Build projects around motivated individuals. Give them the environment and support they need and trust them to get the job done

Remote Adaptation:

- Focus on output, not hours worked. Measure results, not online presence.

- Provide flexible working hours, autonomy over schedules, and home office stipends or tech support.

- Create clear role expectations and ownership maps so people know what they're responsible for.

6. The most efficient and effective method of conveying information to and within a development team is face-to-face conversation

Remote Adaptation:

- Replicate real-time interactions with tools like Zoom, Microsoft Teams, or Gather.

- For async teams, encourage video walkthroughs with Loom or screen recordings with voiceovers.

170

- Use shared digital whiteboards (Miro, FigJam) to visualize ideas as you talk.

Remote Adaptation:

- Use automated tests and CI/CD tools to validate that the software actually works.

- Share sprint demos or project milestones visually with stakeholders.

- Build shared dashboards or public status boards, so progress is always visible.

8. Agile processes promote sustainable development. The sponsors, developers, and users should be able to maintain a constant pace indefinitely
Remote Adaptation:

- Protect deep work time by minimizing meetings and creating no-meeting zones.

- Encourage work/life boundaries with tools like Clockwise or Google Calendar's Focus Time.

- Monitor team health with anonymous pulse surveys or weekly check-ins.

9. Continuous attention to technical excellence and good design enhances agility
Remote Adaptation:

- Encourage regular code reviews using GitHub or Bitbucket.

- Use internal wikis (Notion, Confluence) to document patterns, practices, and decisions.

- Host remote architecture discussions using virtual whiteboards and scheduled "tech talks."

10. Simplicity – the art of maximizing the amount of work not done – is essential

Remote Adaptation:

- Keep sprint goals laser-focused. Limit work-in-progress using digital Kanban boards.

- Regularly audit backlogs to remove stale or unnecessary items.

- Encourage MVP thinking and validate ideas early with stakeholders using short prototypes or mockups.

11. The best architectures, requirements, and designs emerge from self-organizing teams

Remote Adaptation:

- Give teams autonomy to pick their own tools, workflows, and rituals.

- Encourage cross-functional ownership of goals and processes.

- Use async suggestion boxes or discussion threads to gather ideas from the whole team.

12. At regular intervals, the team reflects on how to become more effective, then tunes and adjusts its behavior accordingly

Remote Adaptation:

- Schedule retrospectives regularly (every 2–4 weeks) using tools like FunRetro, MetroRetro, or Miro.

- Allow both live and async reflection options to accommodate time zones.

- Track action items from retros in your project tool and revisit them at the next meeting.

More than just a framework, Agile is a philosophy rooted in adaptability, communication, and continuous improvement. These values are more important than ever in a remote context. By translating Agile's 12 principles into remote-friendly habits and tools, you create a flexible, empowered team culture that thrives even across time zones.

The secret to remote Agile isn't replicating the office virtually. It's designing new rituals and systems that honor the spirit of Agile while respecting the realities of remote work.

Running Remote Sprints, Stand-Ups, and Retrospectives

Agile thrives on rhythm: short, structured cycles of planning, building, reviewing, and improving. For remote teams, maintaining that rhythm can be challenging. Without shared walls, sticky notes, or spontaneous desk conversations, your process must be deliberately designed and digitally facilitated. But here's the good news: remote Agile ceremonies, when done right, can be even more efficient, inclusive, and focused than their in-office counterparts.

In this section, we'll break down how to run remote sprints, stand-ups, and retrospectives, complete with real-world examples, tool recommendations, and adjustments to make them thrive in a distributed environment.

Remote Sprint Planning

Sprint planning sets the tone for everything that follows. It's your team's opportunity to align on what work will be done, why it matters, and how it will be achieved.

Key Elements of Remote Sprint Planning:

- Clear sprint goal

- Prioritized backlog

- Defined capacity (per person and per team)

- Shared agreement on what's "done."

Tools to Use: Jira, ClickUp, Trello, Asana, Notion, Miro

Example Scenario:

A distributed product team in New York, London, and Bangalore uses Notion for backlog grooming and ClickUp for sprint planning. Prior to the meeting, the PM shares a pre-recorded Loom video, walking through the top 10 user stories and tagging each with proposed priority. Team members add time estimates and comments asynchronously.

Then, during a 60-minute live sprint planning call (scheduled during overlap hours), they finalize story selection, confirm capacity, and commit to a clear sprint goal: *"Improve user onboarding by reducing time-to-first-action by 30%."*

Best Practices:

- Share agenda and backlog at least 24 hours in advance

- Use a shared timer during live planning to keep pacing tight

- Assign a notetaker or use an AI meeting assistant to summarize outcomes

Remote Daily Stand-Ups

Stand-ups are the pulse check of Agile. Done well, they build momentum, uncover blockers, and foster team connection. Remotely, they can be live (video call) or async (written or recorded updates).

Formats:

- **Live:** 15-minute video huddle

- **Async:** Slack thread, Trello comments, or a daily check-in form

Structure (same as in-office):

1. What did I do yesterday?

2. What am I working on today?

3. What's blocking me?

Example Scenario – Live:

The design team holds a 10-minute daily Zoom call at 9:00 AM EST. Each person shares their update using the 3-question format. Blockers are noted in a shared ClickUp doc and assigned follow-up owners.

Example Scenario – Async:

A global marketing team uses a Slack bot (like Geekbot or Standuply) to prompt members in their local morning time. The bot compiles responses into a channel where everyone can read updates and respond with emojis or comments.

Best Practices:

- Keep live stand-ups fast and focused (no deep dives – save those for separate chats)

- Encourage camera-optional policy to reduce fatigue

- Review async updates at the end of the day if needed

- Rotate facilitator weekly to build shared ownership

Remote Sprint Reviews / Demos

Sprint reviews (also called demos) give the team and stakeholders a chance to see what's been built and provide feedback.

Key Goals:

- Celebrate progress

- Demonstrate working increments

- Invite input from stakeholders

Formats:

- Live review meeting (Zoom, Teams)

- Async demo board (Loom videos, shared doc with screenshots)

Example Scenario:

A startup's product team hosts a live Zoom demo every other Friday. Engineers walkthrough features they've shipped using a shared screen and take questions in the chat. Meanwhile, the PM records the session, summarizes highlights in Notion, and shares action items.

Alternative Scenario:

A remote QA team uploads short Loom videos showing the new testing dashboard. These are linked in Jira tickets, and stakeholders can comment directly.

Best Practices:

- Keep demos short and structured (2-3 minutes per presenter)

- Record live sessions for those in other time zones

- Use a shared form to collect stakeholder feedback post-demo

Retrospectives are where the magic of continuous improvement happens. Done regularly, they help teams reflect on what's working, what's not, and how to evolve.

Popular Formats:

- Start / Stop / Continue

- What Went Well / What Didn't / What To Try

- Sailboat (anchors, wind, rocks, destination)

Tools: MetroRetro, FunRetro, Miro, Parabol, Google Jamboard

Example Scenario – Live Retro:

Every two weeks, the engineering team logs into Miro and drops sticky notes anonymously. The facilitator clusters topics, the team discusses, and they vote on 1-2 action items to prioritize the next sprint.

Example Scenario – Async Retro:

A cross-functional team spanning six time zones uses Parabol. The tool allows team members to submit feedback asynchronously, cluster themes, and comment on solutions. The PM closes the loop by sharing retro outcomes in Slack.

Best Practices:

- Choose formats that fit your team culture (fun vs. structured)

- Always assign owners and deadlines to action items

- Track actions in your task tool and revisit them in the next retro

- Mix it up! Use different retro formats to keep things fresh

- **Use recurring calendar invites** to anchor team rituals.

- **Automate reminders** with Slack bots or calendar nudges.

- **Start with a check-in question** ("What emoji describes your week?" or "What's one thing you're proud of?") to foster team connection.

- **Protect time zones** by alternating meeting times or recording sessions for fairness.

Remote Agile teams can run sprints, stand-ups, and retrospectives just as effectively — if not more so — than co-located teams. The key is intentionality: clear structure, the right tools, inclusive facilitation, and a culture of continuous feedback.

These ceremonies aren't just boxes to check. They're your team's heartbeat, a way to align, adapt, and improve, no matter where in the world each member logs in from.

Case Studies of Successful Remote Agile Teams

As remote work becomes increasingly prevalent, many organizations have successfully adapted Agile methodologies to distributed environments. Below are three real-world examples of companies that have effectively implemented remote Agile practices.

GitLab: Pioneering a Fully Remote Agile Model

Overview:

GitLab, a DevOps platform, operates with a fully remote workforce spread across more than 65 countries. From its inception, GitLab embraced a remote-first approach, embedding Agile principles into its operations.[1]

Agile Practices in Action:

- **Asynchronous Communication:** GitLab emphasizes asynchronous communication, allowing team members to collaborate across time zones without the need for real-time meetings.

- **Comprehensive Documentation:** The company maintains a publicly accessible handbook detailing its workflows, policies, and best practices, fostering transparency and consistency.

- **Iterative Development:** By breaking down projects into smaller, manageable tasks, GitLab enables continuous integration and delivery, aligning with Agile's iterative approach.

Impact:

GitLab's remote Agile model has led to increased employee satisfaction, higher productivity, and the ability to attract top talent globally. Their success demonstrates

---

[1] https://allmeld.com/article/676-13-case-studies-of-companies-that-successfully-implemented-remote-work

that with the right practices and tools, Agile can thrive in a fully remote setting.[2]

Overview:

During the COVID-19 pandemic, Synchrony Financial faced the challenge of supporting employees in balancing work and childcare. In response, the company launched a virtual summer camp for employees' children within three weeks.[3]

Agile Practices in Action:

- **Cross-Functional Teams:** Synchrony assembled teams from various departments to design and implement the program swiftly.

- **Minimum Viable Product (MVP):** The initial version of the camp was launched quickly, with ongoing iterations based on employee feedback.

- **Continuous Improvement:** The program evolved to include recorded sessions, non-screen activities, and materials for family projects, demonstrating Agile's emphasis on adaptability.

Impact:

The initiative not only alleviated the burden on working parents but also boosted employee morale and

---

[2] https://realitypathing.com/6-case-studies-highlighting-the-future-of-remote-work/
[3] https://time.com/5935054/synchrony-summer-camp/

productivity. It exemplifies how Agile principles can be applied beyond software development to address organizational challenges effectively.

Zapier: Scaling Agile in a Remote-First Environment

Overview:

Zapier, a workflow automation company, has operated as a fully remote organization since its founding in 2011. With over 400 employees worldwide, Zapier has built its operations around remote Agile practices.[4]

Agile Practices in Action:

- **Remote Hiring:** Zapier focuses on hiring individuals who excel in remote work environments, emphasizing self-discipline and communication skills.

- **Regular Check-Ins:** Managers conduct frequent one-on-one meetings to maintain engagement and address any challenges promptly.

- **Outcome-Oriented Metrics:** Performance is measured based on results rather than hours worked, aligning with Agile's focus on delivering value.

Impact:

Zapier's approach has resulted in high employee satisfaction and low turnover rates. Their success

---

[4] https://realitypathing.com/6-case-studies-highlighting-the-future-of-remote-work/

illustrates the effectiveness of integrating Agile methodologies within a remote-first company culture.

These case studies demonstrate that Agile principles can be successfully adapted to remote work environments. By emphasizing communication, flexibility, and continuous improvement, organizations like GitLab, Synchrony Financial, and Zapier have navigated the challenges of remote collaboration and achieved notable success.

Agile Anywhere – Leading with Flexibility and Focus

Agile was never about the room – it was always about the rhythm. While the traditional view of Agile may involve post-it notes on a wall and stand-ups around a whiteboard, today's most successful teams have proven that the core values of Agile can flourish just as well – if not better – across Zoom calls, digital backlogs, and globally distributed teams.

In this chapter, we explored how to adapt Agile's foundational principles to remote and hybrid work environments. Whether it's fostering asynchronous collaboration, building trust in self-organizing teams, or delivering value in rapid, focused increments, Agile provides the mindset and structure remote teams need to stay aligned and responsive. By rethinking rituals like sprint planning, daily stand-ups, and retrospectives through the lens of flexibility and intentional design, teams can retain the spirit of Agile while embracing the realities of digital work.

We also saw how companies like GitLab, Zapier, and Synchrony Financial are walking the talk – leveraging Agile to ship better products, support their people, and stay resilient in the face of complexity. Their successes show that it's not just possible to lead Agile teams remotely – it's often an opportunity to make them even more empowered and outcome-focused.

More than a framework, Agile is a way of thinking and working that values people, progress, and practical results over process for process's sake. And that makes it one of the most powerful tools in your remote leadership toolkit. When practiced with clarity, consistency, and a willingness to adapt, Agile becomes more than a method – it becomes a mindset that connects your team, wherever they are, with a shared purpose and a path to success.

Chapter 12: The Future of Remote Project Management

The future of work isn't on the horizon – it's already here. In just a few years, remote project management has gone from a niche skill set to a fundamental requirement in the modern workplace. However, as tools evolve, team structures shift, and expectations change, project leaders must do more than keep up. They must anticipate, adapt, and lead into what's next.

This chapter explores the emerging trends, technologies, and practices shaping the next generation of remote project management. We'll start with a look at the most significant forces reshaping the landscape, from AI and automation to global hiring, digital-first collaboration, and employee well-being. You'll learn what's coming, what's sticking, and what's being left behind.

Next, we'll tackle one of the most talked-about shifts in modern work: hybrid models. While some teams stay fully remote and others return to the office, many organizations are navigating the in-between. We'll discuss how to design hybrid setups that are equitable, efficient, and purpose-driven – avoiding the pitfalls of "two-tier" cultures or meetings with half the team dialed in from a coffee shop.

Finally, we'll close the chapter by bringing the focus back to you: the remote project manager. Success in this evolving landscape depends on continuous improvement, not just in processes and tools, but in your mindset, leadership, and adaptability. You'll leave this chapter with practical strategies to stay future-ready, keep your skills sharp, and lead with confidence wherever the work takes you.

Because the future isn't remote or in-office, it's flexible, dynamic, and human-centered, and project managers who can lead in that space will be the ones driving meaningful change.

Trends Shaping the Future of Remote Work

Remote work is no longer a temporary experiment but a defining feature of modern work culture. But just because we've adjusted to Zoom calls and home offices doesn't mean the story ends there. The remote landscape continues to evolve rapidly, driven by technological innovation, shifting workforce expectations, and global socio-economic forces.

For remote project managers, staying ahead of these trends is essential. The ability to lead distributed teams will increasingly depend on understanding the forces shaping this new world of work. Here are the key trends shaping the future of remote project management and what they mean for you.

1. AI and Automation Everywhere

From scheduling meetings to generating sprint summaries, AI is transforming how remote teams operate. Tools like ChatGPT, Notion AI, Fireflies, and ClickUp AI are reducing administrative overhead and providing real-time insights into team performance, risks, and workload.

Implications for PMs:

- Focus less on reporting and more on strategic decision-making

- Learn how to evaluate and deploy AI tools responsibly

- Stay aware of ethical and data privacy concerns

2. The Rise of Async-First Cultures

As time zones stretch across continents, asynchronous work has become more than a convenience; it's a necessity. Teams are embracing async stand-ups, decision logs, pre-recorded demos, and written feedback instead of live meetings.

Implications for PMs:

- Master async communication tools (Loom, Slack, Notion)

- Set clear expectations and communication protocols

- Create structured documentation habits

3. Work-Life Integration and Mental Health Prioritization

Remote work has blurred the boundaries between personal and professional life. Teams are demanding more flexibility, not just in where they work but how and when they work. At the same time, burnout, loneliness, and mental health concerns are on the rise.

Implications for PMs:

- Foster a culture that respects offline time and autonomy

- Normalize mental health check-ins during retros or 1:1s

- Encourage work habits that support sustainability (e.g., no-meeting days, async Fridays)

4. Global Talent, Local Challenges

Remote work opens access to a global talent pool. However, it also introduces complexity in managing cultural differences, legal compliance, and time zone coverage.

Implications for PMs:

- Develop cross-cultural communication and leadership skills

- Balance inclusivity with efficiency (e.g., rotate meeting times, use translations where needed)

- Partner with HR or EOR (Employer of Record) platforms to stay compliant with labor laws

5. Outcome-Based Performance Management

With remote teams, visibility doesn't come from who's online – it comes from what's delivered. Companies are shifting away from time-based tracking toward outcomes and impact.

Implications for PMs:

- Align teams around clear, measurable goals (OKRs, SMART goals)

- Use dashboards to track deliverables and progress in real-time

- Focus performance reviews on value created, not time-logged

While some companies remain fully remote and others return to the office, most are building hybrid models. These models aim to combine the flexibility of remote work with the collaboration benefits of co-location, but they're still evolving.

Implications for PMs:

- Build inclusive practices that don't favor in-office visibility over remote contributions

- Equip teams with tools and rituals that work across formats (e.g., remote-friendly meeting design, co-located + dial-in setup)

- Help define the "why" for in-person time (e.g., team offsites, deep planning, innovation sprints)

7. Increased Focus on Security and Digital Literacy

As remote teams share sensitive data across tools and borders, cybersecurity becomes critical. Similarly, every team member needs to be fluent in using remote-first tools.

Implications for PMs:

- Partner with IT to ensure secure tool usage and data policies

- Provide training and SOPs for tool usage (don't assume digital fluency)

189

- Create contingency plans for outages or tech failures

8. Project Managers as Coaches, Not Controllers

With more autonomy comes a shift in leadership style. Project managers are increasingly expected to guide and coach, not micromanage. The future PM is less of a taskmaster and more of a servant-leader.

Implications for PMs:

- Build emotional intelligence, active listening, and feedback skills

- Focus on enabling teams to self-organize and solve problems

- Create psychological safety by modeling vulnerability and transparency

9. The Virtual Office is Getting Smarter

From VR workspaces to digital HQs like Gather and Teamflow, remote workspaces are evolving beyond simple video calls. These platforms aim to replicate the social and spatial awareness of the office in virtual form.

Implications for PMs:

- Experiment with digital co-working spaces for collaboration or casual connection

- Explore tools that support visual planning, spatial awareness, and presence

- Monitor what adds real value vs. novelty; don't overload your stack

The future of remote work is not a straight line; it's a dynamic, evolving ecosystem. As a project manager, your role is to observe, adapt, and lead through change. Stay curious. Be ready to experiment. And above all, keep people at the center of your process.

Hybrid Work Models: Finding the Right Balance

As companies navigate the future of work, hybrid models have emerged as the most common middle ground between fully remote and fully in-office setups. But make no mistake, hybrid is not a compromise. When done intentionally, it can be the best of both worlds: providing flexibility for individual focus and connection for collaborative moments.

That said, hybrid isn't just about who works where. It's about designing workflows, communication, and culture that are equitable for everyone, regardless of location. Without thoughtful planning, hybrid can quickly become two different experiences under the same company roof — creating silos, resentment, and lost opportunities.

Here's how project managers can help shape hybrid work models that actually work.

Define the Why of In-Person Time

Avoid falling into the trap of arbitrary office mandates. Instead, clarify the value of in-person collaboration:

- Strategic planning and visioning

- Relationship-building and social connection

- Cross-functional innovation sprints

- Onboarding and mentorship

Design co-located moments around intentional outcomes and not just presence.

Normalize Remote Participation as the Default

Even if some people are in the same room, always structure meetings as if everyone were remote. This ensures full inclusion and prevents remote attendees from being sidelined.

Best Practices:

- Everyone dials in separately (even in-office attendees)

- Use shared docs and digital whiteboards instead of physical ones

- Record and summarize meetings for those not present

Invest in Asynchronous Collaboration

Hybrid success relies on mastering async work. If progress depends on syncing schedules every day, the model breaks down.

Tips:

- Use tools like Loom, Notion, and Slack for async updates

- Create shared project spaces with all relevant info and context

- Establish team agreements on response times, availability, and handoffs

Design for Equity, Not Uniformity

Not everyone needs to work the same number of days in-office, but everyone should have equal access to opportunity.

Equity Questions to Ask:

- Are promotions and high-visibility projects distributed fairly?

- Do remote team members get facetime with leadership?

- Are cultural events and celebrations accessible to all?

Create Clear, Flexible Policies

Hybrid thrives on clarity and autonomy. Document your policies in a living team handbook that answers:

- How often (and why) people should be in person

- What tools are used for communication and collaboration

- How decisions are made and documented

Avoid vague expectations, as ambiguity leads to inconsistency.

Rethink Office Design and Purpose

When teams do come together, the office should feel like a tool, not a tether.

- Replace rows of desks with collaborative spaces

- Offer reservable meeting rooms and hot desks

- Support "drop-in" days with snacks, tech support, and team rituals

Make in-person time feel valuable, not obligatory.

Track the success of your hybrid strategy with real data and not just assumptions.

- Survey employee satisfaction and sense of belonging

- Track team productivity and delivery cycles

- Monitor meeting engagement and tool usage

Use what you learn to iterate your hybrid approach over time.

Hybrid isn't a one-size-fits-all solution. It is a framework that needs to be customized for your team's goals, culture, and workflows. As a project manager, your job is to design clarity, promote equity, and enable momentum, whether your team is spread across floors or continents.

With the right balance of structure and flexibility, hybrid models can give teams the freedom they crave and the connection they need.

Continuous Improvement as a Remote Project Manager

In remote project management, improvement isn't a one-time initiative but a way of working. As technology shifts, teams evolve, and work culture changes, the best project managers are those who never stop learning, adjusting, and growing. Continuous improvement is both a mindset and a strategy, and in a remote context, it's your key to staying resilient, relevant, and ready for anything.

Here's how to build a rhythm of growth and improvement into your daily leadership practice.

1. Retrospectives Aren't Just for Teams

Most Agile teams hold regular retrospectives to evaluate what's working and what's not, but PMs often forget to hold retrospectives for themselves.

Make it a habit to review your own performance:

- What went well in the last sprint or project?

- What could have been communicated more clearly?

- Were timelines realistic? Were goals aligned?

Even better, ask your team for feedback. A short, anonymous survey after each project or major milestone can reveal blind spots and highlight areas for growth.

Tool Tip: Use tools like Google Forms, Officevibe, or Polly in Slack for lightweight check-ins.

2. Develop Your Remote-Specific Leadership Skills

Remote leadership isn't just about managing tasks but is also about enabling people to thrive in autonomy, ambiguity, and distance. That requires a unique blend of skills:

- **Emotional intelligence**: Reading between the lines in messages or Zoom calls

- **Clarity and documentation**: Translating vision into asynchronous, written plans

- **Digital facilitation**: Running engaging, inclusive meetings and workshops

- **Cultural awareness**: Navigating time zones, holidays, and communication styles

Invest in your skills the same way you encourage your team to develop theirs. Read widely. Attend webinars. Join communities of practice like Remote Work Association, PMI, or LeadDev.

3. Track What You Want to Improve

You can't improve what you don't measure. Choose a few key areas to focus on each quarter and build a habit of reviewing them.

Examples:

- % of projects delivered on time

- Team engagement scores

- Average meeting length (and satisfaction)

- Frequency of unplanned scope changes

- Your own work/life balance or time spent in deep work

These metrics don't need to be perfect – but they give you a baseline to grow from.

4. Build a Feedback-Rich Culture

Feedback isn't just for performance reviews! It's how teams and leaders stay aligned and adaptable. Model regular feedback by asking for it, giving it, and acting on it.

- Start 1:1s with "What's one thing I could do better as your PM?"

- End retros with "How did this project go from a leadership perspective?"

- Use "start/stop/continue" formats to make feedback easy and actionable

When people see that feedback leads to change, they'll be more likely to offer it, and your leadership will evolve to meet their needs.

5. Share What You're Learning

Don't just learn, but share your learning with others. Whether it's a new facilitation trick, a better tool for async stand-ups, or a case study you read, bring that back to your team.

- Create a "PM Learnings" Slack channel

- Write monthly reflection posts on your intranet or in Notion

- Host quarterly "what we're improving" sessions

Transparency breeds trust, and sharing lessons helps your whole team grow – not just you.

6. Protect Time for Strategic Thinking

It's easy to get stuck in execution mode: checking in on tasks, updating timelines, and responding to Slack. But continuous improvement also means stepping back to ask, *Are we doing the right things the right way?*

Block time each week or month to reflect on:

- Are our processes still serving us?

- Are we aligned with our goals?

- Where are the friction points in how we work?

Use this time to revise workflows, sunset outdated rituals, or pilot small experiments. Improvement isn't always about big overhauls. Often, it's the 1% tweaks that change everything.

Being a great remote project manager isn't about having all the answers. It's about being committed to asking better questions, experimenting with new approaches, and evolving alongside your team.

Continuous improvement isn't just something you *do* – it's a way you *lead*. Stay curious, stay open, and never stop building better ways to work. Because the future will keep changing, and with the right mindset, you'll be ready to lead it.

Conclusion & Final Takeaways

As we reach the end of this journey into the world of remote project management, one thing should be clear: remote isn't just a location. It is a mindset, a skillset, and a new standard of leadership. Over the course of this book, we've explored the practical strategies, tools, and habits that help project managers lead with clarity, confidence, and agility no matter where their teams are located.

Recap of Key Lessons

Let's take a moment to distill the most important lessons we've learned. More than just summaries, these lessons are your remote project management compass, guiding your decisions and reinforcing the core principles of effective, human-centered leadership in a distributed world.

1. The Remote Landscape Has Changed, And So Must You

Remote and hybrid work are no longer exceptions. They are embedded in the DNA of modern teams. To thrive, project managers must evolve from process enforcers to adaptive leaders. That means learning to lead without proximity, building trust without face time, and delivering outcomes without micromanagement. The fundamentals of good project management still apply, but the way they're practiced has changed.

2. Mindset Is Your Foundation

Remote project success begins not with tools or templates but with mindset. The best remote leaders lead with empathy, clarity, and flexibility. They embrace ambiguity, foster trust, and promote autonomy. They don't control every detail. Rather, they create the conditions where teams can thrive, make smart decisions, and do great work independently.

3. Team Setup Is Everything
Building a high-performing virtual team requires intentionality. Hiring the right people is only step one. Onboarding, defining clear roles and expectations, and creating shared agreements around communication and collaboration are just as critical. A well-structured team sets the tone for accountability, trust, and psychological safety.

4. Planning Still Matters, It Just Looks Different
Whether you're using Agile, Scrum, Waterfall, or a hybrid model, project planning in a remote context must be visual, transparent, and collaborative. Goals must be clearly defined, deliverables explicitly assigned, and accountability structures embedded into your tools and rituals. Kickoff meetings, progress tracking, and stakeholder alignment don't disappear in remote work — they become even more important.

5. Communication Is Your Superpower
In a remote environment, communication isn't just a soft skill; it's a survival skill. Knowing when to use synchronous vs. asynchronous communication, which tools to use, and how to prevent miscommunication is essential. Whether it's Slack, Zoom, email, or Notion, how you communicate sets the tone for your team's culture, productivity, and cohesion.

6. Culture Doesn't Happen by Accident

Remote teams can't rely on hallway chats or shared lunches to build camaraderie. Leaders must design for connection through rituals, recognition, team-building activities, and space for human interaction. Culture lives in how you celebrate wins, handle conflict, and support one another.

7. Productivity Is About Outcomes, Not Hours

Tracking performance in remote teams requires a shift from activity to impact. SMART goals, transparent workflows, and regular feedback loops help ensure that everyone stays aligned. It's not about who's online the longest; it's about who's delivering value. And when distractions or time zone differences arise, systems – not supervision – solve the problem.

8. Risk and Resilience Go Hand in Hand

Remote work introduces unique risks, from communication breakdowns to tech outages to disengaged team members. But with the right contingency planning, visibility, and check-ins, you can mitigate these challenges and lead with resilience. Being calm, transparent, and solution-oriented when things go wrong is one of your greatest assets.

9. Meetings Must Be Intentional

Remote meetings shouldn't be a default; they should be a last resort. When necessary, they must be short, focused, and structured. Tools like recordings, AI summaries, and shared agendas help make meetings more efficient, inclusive, and less fatiguing. And many traditional meetings can be replaced with well-crafted async updates.

10. Technology Is Your Teammate

The right tech stack doesn't just support your workflow, but rather, it powers it. AI and automation tools reduce friction, streamline communication, and support faster decision-making. Integrated platforms like Jira, Asana, ClickUp, or Trello keep everyone aligned. And cybersecurity practices ensure that trust and data stay protected.

11. Agile Can Work Anywhere

Agile isn't tied to a physical board or daily huddle. It is a mindset and method that translates beautifully to remote and hybrid teams. Whether it's running remote stand-ups, retrospectives, or sprint planning sessions, Agile gives teams structure without rigidity. And real-world case studies have shown that distributed Agile teams can outperform their co-located peers when led with intentionality.

12. The Future Is Hybrid, Flexible, and Human-Centered

The future of remote project management isn't just about remote but about choice, balance, and purpose. Hybrid models, asynchronous-first cultures, and outcome-based work are here to stay. The most effective leaders will be those who understand how to design systems that support both flexibility and performance while keeping people at the heart of the process.

13. Your Growth Is the Team's Growth

The most important tool in any remote leader's toolkit is not a platform or process but a mindset of continuous improvement. By investing in your own learning, seeking feedback, and modeling vulnerability, you create a team culture that grows, adapts, and innovates together. Leadership isn't about knowing everything – it's about getting better together.

These lessons are actionable. They're the foundation for managing projects in a remote world with clarity, creativity, and confidence. In the next section, we'll challenge you to put what you've learned into motion because the best way to learn remote leadership is to live it, one smart, strategic action at a time.

A Personal Challenge: Taking Action Today

By now, you've read through frameworks, tools, case studies, and strategies for managing projects in a remote world. You've seen what works, what doesn't, and how to adapt your leadership to meet the demands of distributed teams. But here's the truth: none of this matters if it stays on the page.

So, here's your challenge: **take one meaningful action today.**

Not tomorrow. Not when your next project kicks off. Today.

You don't need to implement a full remote strategy overhaul. You don't need to set up a dozen new tools. You just need to begin. The difference between a good leader and a great one often lies in momentum and in taking the first step with purpose.

Here are a few ways to start:

- **Audit your next meeting**: Is it necessary? Can it be async? Could you improve it with a shared agenda or a clearer outcome?

- **Reach out to your team**: Schedule a 1:1 with someone you haven't connected with in a while. Ask them what's working and what's not.

- **Document one process**: Choose something you've been explaining repeatedly (onboarding, sprint planning, retros) and write it down. Share it.

- **Start a feedback loop**: Send a short, anonymous survey or Slack poll asking for one way you could improve as a remote leader.

- **Pick one tool to learn better**: Whether it's Notion, Jira, Loom, or Slack, choose a platform you use daily and explore a feature you haven't mastered.

It doesn't matter what you choose. What matters is that you choose *something*. Because remote leadership is a craft. And like any craft, it sharpens with use, not just study.

One action today leads to another tomorrow. One documented process becomes a shared playbook. One open conversation builds more trust. And before long, you'll look back and see that you didn't just read a book on remote project management – you became the kind of leader this book was written for.

Start small. Stay consistent. Lead forward.

Resources for Continued Learning

Your journey as a remote project manager doesn't end here. To further enhance your skills and stay updated with evolving practices, consider exploring the following resources:

Books

- **Remote: Office Not Required** by Jason Fried & David Heinemeier Hansson

  A foundational guide on building and managing remote teams effectively

- **The Long-Distance Leader: Rules for Remarkable Remote Leadership** by Kevin Eikenberry & Wayne Turmel

-

  Offers practical advice on leading remote teams with clarity and purpose.

- **Leading from Anywhere: The Essential Guide to Managing Remote Teams** by David Burkus

  Provides actionable strategies for remote leadership and team engagement.

- **The Remote Project Manager** by Gren Gale

  Focuses on tools and soft skills required for successful remote project delivery.

Podcasts

- **The Project Management Podcast** hosted by Cornelius Fichtner offers insights from global PM experts.

- **Projectified Podcast** produced by PMI discusses emerging trends and challenges in project management.

- **Project Management Happy Hour**, Co-hosted by Kim Essendrup, tackles real-world PM challenges with humor and practical advice.

- **The Digital Project Manager Podcast** Focuses on digital project management strategies and tools.

Online Courses & Certifications

- **PMI Kickoff™**

  A free, 45-minute digital course introducing the basics of project management. https://www.pmi.org/learning/free-online-courses

- **Google Project Management Certificate (Coursera)**

  Offers comprehensive training in project management, including AI applications. https://grow.google/certificates/project-management/

- **Coursera Project Management Courses**

  Features courses from institutions like the University of California, Irvine, and IBM, covering various PM methodologies.

https://www.coursera.org/courses

- **Remote Projects 101: The Remote Guide to Project Management**

  A comprehensive guide covering each stage of remote project management, challenges, and best practices.

- **Smartsheet's Remote Project Management E-Book**

  Provides insights into the top challenges facing remote project managers and actionable tips to overcome them.

By engaging with these resources, you can continue to refine your skills, stay abreast of industry trends, and lead your remote projects with confidence and competence.

www.ingramcontent.com/pod-product-compliance
Lightning Source LLC
Chambersburg PA
CBHW071557210326
41597CB00019B/3291